THE GERMAN AMERICAN
FAMILY ALBUM

THE GERMAN AMERICAN
FAMILY
ALBUM

DOROTHY AND THOMAS HOOBLER
Introduction by Werner Klemperer

OXFORD UNIVERSITY PRESS • NEW YORK • OXFORD

Oxford University Press

Oxford New York
Athens Auckland Bangkok Bombay
Calcutta Cape Town Dar es Salaam Delhi
Florence Hong Kong Istanbul Karachi
Kuala Lumpur Madras Madrid Melbourne
Mexico City Nairobi Paris Singapore
Taipei Tokyo Toronto

and associated companies in
Berlin Ibadan

Design: Sandy Kaufman
Layout: Valerie Sauers
Consultant: Sally M. Miller, Professor of History, University of the Pacific, Stockton, California

Published by Oxford University Press, Inc.,
198 Madison Avenue, New York, New York 10016

Oxford is a registered trademark of Oxford University Press

Library of Congress Cataloging-in-Publication Data

Hoobler, Dorothy.
The German American family album / Dorothy and Thomas Hoobler; introduction by Werner Klemperer
p. cm. — (American family albums)
Includes bibliographical references and index.
1. German American families—Juvenile literature. 2. German Americans—History—Juvenile literature.
I. Hoobler, Thomas. II. Title. III. Series.
E184.G3H65   1995
973'.0431—dc20   95-14448
                                                       CIP
                                                       AC

ISBN 0-19-508133-1 (lib. ed.); ISBN 0-19-510341-6 (trade ed.); ISBN 0-19-510172-3 (series, lib. ed.)

1 3 5 7 9 8 6 4 2

Printed in the United States of America
on acid-free paper

Cover: The family of Adeline and Heinol Staffel at their golden wedding anniversary
around 1900 in San Antonio, Texas.

Frontispiece: A family in New Ulm, Minnesota, around 1900.

CONTENTS

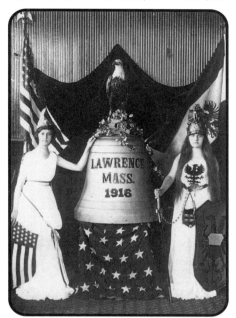

"Liberty and Germania," a patriotic display in 1916, linked German and American symbols.

INTRODUCTION

by Werner Klemperer

As a country, Germany has given us an incredible artistic heritage—especially in music. My family was very much a part of that tradition. My father, Otto Klemperer, was an internationally known conductor who performed all over Europe and in America, too. He was born a Jew, though he converted to Roman Catholicism at the age of 22. (He returned to Judaism in 1967, six years before he died.) My mother, Johanna Geissler, was an opera singer and a leading soprano at the Cologne Opera, where she met my father. Three years later, they were married. My sister and I were raised as Catholics, but that did not save us from the devastation of the Nazi regime.

I was born in Cologne, and my family moved very shortly thereafter to Wiesbaden, where my father was appointed head of the opera house. We lived there for three years, until my father received an offer from Berlin, so we moved there, where we lived until we immigrated.

A few months before we left Germany, my father took me out of school because there had been some trouble for some of the Jewish students. We departed from Germany rather suddenly. My father was in Switzerland and he called my mother and said that he wanted us all out of Germany as quickly as possible. We took a train for Switzerland a few days later. It was very early, really, in the scope of German Jewish emigration.

There was no real drama to our trip except for one story. My mother brought a cake for us to eat while we were on the train. It was one of those cakes about five or six inches high, with a hole in the center. As we got close to the border, she ordered something to drink—coffee or tea, I think—with the cake. She sliced a few pieces, and we were eating cake when the border patrol came in and checked the compartment and our suitcases. This inspection was done on all the trains in those days. Years later, I found out that my mother had baked into the cake a hundred thousand marks and that way smuggled it out of Germany. That took a lot of guts on my mother's part.

We lived in Switzerland for six months and then moved to Vienna for almost two years. In June 1935, we arranged to come to the United States. When we arrived in New York, we took the train to Los Angeles, where my father had accepted the post of musical director of the Los Angeles Philharmonic.

In Germany my sister and I had had cursory lessons in English, but we knew very little of the language. We were having a difficult time. To solve the problem, my father had a very ingenious idea. He hired a graduate student at UCLA to be our tutor, but the tutor didn't understand or speak a word of German. We were with him from nine in the morning until three or four in

Werner and his sister, Lotte, with their mother, opera singer Johanna Geissler, in Berlin around 1930.

World-renowned conductor Otto Klemperer rehearses with the Vienna Philharmonic in the late 1950s.

The Klemperer family in Los Angeles in 1938.

Klemperer in the U.S. Army, 1943.

Werner Klemperer, whose career encompasses theater, motion pictures, television, and the concert stage, was born in Germany and emigrated to the United States while still a child with his father, Otto Klemperer, the illustrious symphony and opera conductor. He is best known to audiences worldwide for his starring role as Colonel Klink in the classic TV series "Hogan's Heroes," for which he won two Emmy awards. TV audiences continue to enjoy his work, most recently on "Law and Order" and "The Simpsons." He is a recipient of the Disney Studio Legend of Television Award.

Klemperer's distinguished film career includes classics such as Ship of Fools, Judgment at Nuremberg, The Goddess, Houseboat, and The Wrong Man. His numerous stage credits include The Sound of Music, The Master Class, Idiot's Delight, Cyrano de Bergerac, The Merry Widow, and the Broadway revivals of Cabaret (Tony award nomination) and Uncle Vanya.

Klemperer in Uncle Vanya in 1995.

On the concert stage Klemperer is considered the foremost narrator of the classical orchestra repertoire. He has been soloist with the New York Philharmonic, Philadelphia Orchestra, Boston Symphony, Chicago Symphony, and most other major orchestras. With the Mostly Mozart Festival Orchestra, he was featured on PBS television's "Live from Lincoln Center."

Klemperer recently released a recording of the Dr. Seuss story Gerald McBoing Boing, and he has narrated Babar the Elephant at the annual Christmas party for ambassadors' children at the White House.

the afternoon, and we had to fight our way through words and sign language, but it turned out to be a most fabulous experience. The tutor took us to the beach in the afternoons, and by the time school started in the fall, we had acquired a working knowledge of English. (At home, though, we continued to speak German.)

There was an extensive emigré community in Los Angeles. You would have thought that most of the leading German artists would have emigrated to New York, but for some reason many of them settled on the West Coast, including the writers Thomas Mann, Bertolt Brecht, and Franz Werfel; the composer Arnold Schoenberg (my father was responsible for getting him a job as a professor at UCLA) and Erich Korngold; the conductor Bruno Walter; the film directors Fritz Lang and Otto Preminger—and those are just a few of them.

The German community was fairly close-knit. I don't know whether they all had close relationships, but there always existed a kind of a salon atmosphere. An emigré woman named Salka Viertel, an important editor at MGM and a close friend of Greta Garbo, hosted this salon. She knew everybody. Every Sunday, at Salka's house at the beach, everybody gathered for afternoon coffee and tea. We were just kids, so we were left outside to play while the adults sat around and discussed everything from politics to cultural matters. Here

was a remarkable gathering of refugees who would enrich the cultural life of the United States by so much. Growing up in this stimulating atmosphere must have, in some way, influenced my decision to become an actor.

I have never been a person who has incredible patriotic feelings, although I think it is essential to be a good citizen. My loyalty to Germany has never been political, but culturally, this loyalty is always there. What happened to Germany politically in the 1930s discolored German culture in some bizarre way. When the Nazis took over the country, there was a tremendous break, and mediocrity set in. It is my feeling that some of it still has to be overcome.

There is no question that Germany must be counted as one of the most important countries in terms of hundreds and hundreds of years of great culture that has developed there. Every time I think of the rich culture of the great writers, philosophers, composers, and painters, I am truly proud of my German heritage.

As an American, let me conclude, how grateful I am for the opportunities this country has given me to pursue my profession as an actor.

Werner Klemperer [signature]

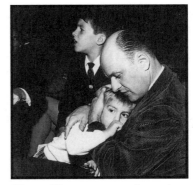

Werner Klemperer at a rehearsal in the early 1970s with his two children, Mark and Erika.

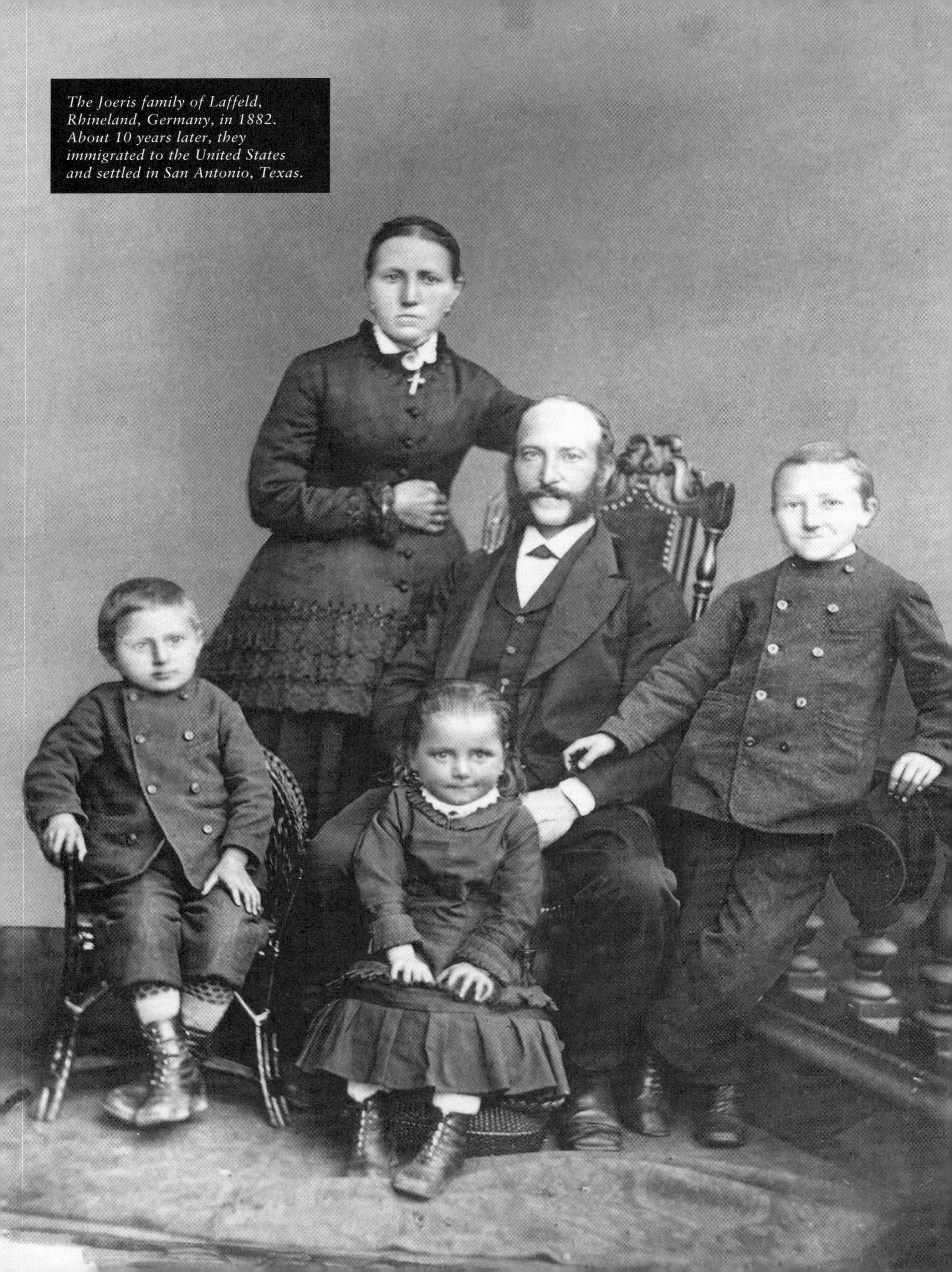

The Joeris family of Laffeld, Rhineland, Germany, in 1882. About 10 years later, they immigrated to the United States and settled in San Antonio, Texas.

CHAPTER ONE

THE OLD COUNTRY

The first German to reach the New World was a sailor named Tyrker, a companion of the Scandinavian seafarer Leif Ericsson. Sometime after the year 1000, Ericsson landed in Labrador, an island off the coast of Canada. According to the Icelandic sagas, when Tyrker went ashore he found grapevines. Thus he called the new land Vinland, or Vineland.

Five centuries later, more European explorers began to cross the Atlantic. One of them, an Italian named Amerigo Vespucci, wrote an account of his voyage that was read by a German mapmaker, Martin Waldseemüller. In 1507, Waldseemüller gave the new continents the name America.

At the time of the early European explorations of America there was no country called Germany. The German-speaking peoples of north-central Europe lived in many different states that were part of a loose confederation called the Holy Roman Empire.

Charles V, a member of the Hapsburg family, was crowned Holy Roman Emperor in 1519. Two years earlier, the German monk Martin Luther had begun the religious Reformation that challenged the authority of the Roman Catholic church.

Lutheranism spread through the German states, winning the support of some of their rulers. A series of wars between Catholic and Lutheran states followed. In 1555, Charles agreed to the Peace of Augsburg, which allowed the ruler of each German state to choose the religion of his subjects.

Other Protestant religions sprang up after the success of Lutheranism. In Switzerland, John Calvin's ideas led to the formation of what was called the Reformed church. After the Peace of Augsburg, members of the Reformed church won converts in some of the German states. Meanwhile, the Roman Catholic church had begun the Counter-Reformation to win back the "heretic" or dissenting Protestants. Tensions increased among Calvinists, Lutherans, and Catholics.

In 1619, Ferdinand II, a devout Catholic, became Holy Roman Emperor. The year before, some of his Protestant subjects had started a rebellion that was supported by the ruler of a German state called the Palatinate. Ferdinand responded to this political and religious challenge by sending troops to crush the rebellion.

This was the beginning of the Thirty Years' War (1618–1648), a long and bloody conflict that eventually drew in most of the other European powers. For three decades, armies of the warring nations marched through the German states, killing not only each other but also the local citizenry, burning crops and villages, and leaving devastation in their wake. About one-third of the population of the German states died, either in the fighting or from the starvation and epidemics that followed. At last, the war came to an end in 1648 with the Peace of Westphalia.

The peace agreement left the German states weaker and more divided than before. There were now more than 300 "Germanies," some no larger than a single city. For more than a century, the German states fell prey to their stronger neighbors, such as Sweden and France.

During this time, some Germans began to seek haven elsewhere. The first German American settlement in today's United States was founded in 1683 in Pennsylvania, and more followed in the next century, when warfare again raged through the German states.

In 1763, Empress Catherine the Great of Russia, who was herself German, offered free land and religious freedom to foreigners who would agree to settle in the southern parts of her empire. Catherine hoped to populate the vast steppe region with loyal citizens who would strengthen her domain. Many Germans accepted the offer. A century later, when the Russian government adopted a hostile policy, sizable numbers of these "Russian Germans" emigrated to the United States.

Meanwhile, Germany began the long process of unification. Under the leadership of several able kings, the state of Prussia built an army that rivaled those of the leading powers of Europe. Throughout the 18th century, Prussia added territory to its domain. By 1772, when Prussia seized parts of Poland, it was the largest and most powerful German state except for Austria.

In spite of the division and warfare that plagued the German people, they developed a rich culture. The writer Johann Wolfgang von Goethe (1749–1832) expressed the German love of nature and the idea that human intellect must be balanced by deep emotion. Goethe influenced other German writers such as the playwright and poet Friedrich von Schiller and the philosopher Immanuel Kant.

German culture had its most enduring influence in music. During the 18th and 19th centuries, the rulers of many German states maintained lavish courts, supported by heavy taxes on their subjects. To add to the splendor of their courts, the rulers sponsored gala musical entertainments. The world benefitted, for a series of German-speaking musical geniuses produced works that have never been surpassed. Starting with

Johann Sebastian Bach (1685–1750), the list runs through Joseph Haydn, Wolfgang Amadeus Mozart, Ludwig van Beethoven, Franz Schubert, Robert Schumann, Richard Wagner, Johann Strauss, and Johannes Brahms (1833–97). Wherever German emigrants settled, they brought with them their glorious musical tradition.

At the beginning of the 19th century, the German states, like the rest of Europe, became engulfed in the Napoleonic Wars. After Napoleon Bonaparte seized power in France in 1799, he em-

A composite picture of the Beiser family in Hamburg, Germany, around 1900. The faces of the two adults at left, Arthur and Agnes, were inserted by the photographer. They had emigrated to the United States.

barked on a series of conquests. Napoleon's forces soundly defeated Austria, bringing the Holy Roman Empire to an end. Prussia also suffered, losing about half its territory. But after Napoleon's disastrous invasion of Russia in 1812, the tide turned against him. A coalition of European powers, including Prussia and Austria, took part in Napoleon's final defeat in 1815.

The victors met at the Congress

of Vienna to redraw the map of Europe. Prussia absorbed territory along the Rhine River, called the Rhineland. Austria became an empire that included not only German-speaking people but also Czechs, Slovaks, and Hungarians.

As a result of the Congress of Vienna, a German Confederation was created, with Austria and Prussia as its leading members. The hundreds of German states were consolidated into only 39. However, this Confederation was weak and never achieved the status of a real government.

The Napoleonic Wars had spurred the spirit of nationalism throughout Europe. Germans eagerly discussed the idea that they should unite into one powerful nation.

However, the various non-Germans in the Austrian Empire wanted their own independence. In the fateful year 1848, revolts broke out all over Europe—in Austria, the smaller German states, and France and Italy.

Some of the revolutionaries, inspired by democratic ideals, hoped to establish constitutional governments. However, all of the revolutions of 1848 were doomed to failure and were cruelly crushed within a year. Many German revolutionaries fled, some to the United States.

The German Confederation was restored, though it was clear that Prussia and Austria would never yield power to each other. In addition, the princes, dukes, margraves (military governors), and other rulers of the smaller states jealously guarded their own authority.

In 1862, Otto von Bismarck became prime minister of Prussia. He was determined to unite the German lands under Prussia's leadership. Though he often had to urge a timid King William I to go along, Bismarck used Prussia's military might to attain his goals. In 1866, Prussia's army defeated the Austrians at the Battle of Sadowa. From then on, Bismarck excluded Austria, with its mixed population, from his plans for a united Germany.

Four years later, in response to a French attack on German territory, Bismarck forged a military alliance of German states. The Franco-Prussian War ended with a decisive victory by Bismarck's forces at the Battle of Sedan in September 1870. Bismarck now used his diplomatic skills to persuade the rulers of the other German states to set aside their differences and unite into a German empire. In January 1871, King William accepted the title Emperor of Germany. For the first time, Germany was one nation.

Bismarck cemented German unification by standardizing the new nation's legal, financial, and governmental systems. He also launched a *Kulturkampf* ("culture struggle") to reduce the influence of the Roman Catholic church in Germany. Some German Catholics responded to Bismarck's policies by emigrating.

In 1890, a new German emperor, William II, forced Bismarck to retire. Only 29 years old when he took the throne, William II had ambitious ideas. He built up his army and navy, preparing for war.

In 1914, World War I broke out in Europe. Germany was allied with Austria against France, Britain, and Russia. For three years, the nations of Europe fought each other in battles made more devastating by new weapons such as machine guns, airplanes, and heavy artillery. In November 1917, the United States entered the war on the side of France, Britain, and Russia. A year later, Germany and its allies were forced to surrender. William II abdicated and Germany became a republic.

A street in Tiergen, Germany, around 1920. Although German cities were untouched by the fighting during World War I, Germany suffered economic hardship in the years that followed.

At the Peace Conference at Versailles, France, in 1919, the victors imposed harsh terms. The new German government was required to give up part of its territory to pay huge reparations to the victorious nations, and to drastically reduce its army and navy.

Germany's people resented the humiliation and economic hardship caused by the war reparations. Post-war inflation made their paper money almost valueless. Capitalizing on widespread unrest, a former corporal in the German army named Adolf Hitler began to attract support for his National Socialist (Nazi) Party, which pledged to restore Germany to its former greatness. In 1933, the Nazis won enough seats in the Reichstag (Parliament) to make Hitler the chancellor. He soon was given dictatorial powers and set out on a policy of aggression.

In 1939, Hitler's troops invaded Poland, setting off the conflict that became known as World War II. The German army won a series of victories that put Hitler in control of most of the European mainland by 1944.

Blaming Jews for Germany's economic woes, Hitler herded them into camps to be used as slave laborers or killed. The German Jewish community, which had contributed to German life and culture since the Middle Ages, was virtually destroyed in the Holocaust.

However, Hitler squandered his military resources by invading the Soviet Union. In June 1944, a combined American and British force landed in Nazi-controlled France. Soviet troops invaded Germany from the east. Within a year, Hitler's regime lay in ruins.

After the war, Germany was divided into four zones, administered by the United States, Britain, France, and the Soviet Union. Eventually, the Soviet zone became Communist-dominated East Germany, the other three zones the democratic country of West Germany. In 1990, with the collapse of communism, East and West Germany combined into a single nation once again.

The wars of the 20th century have shifted Germany's borders. The former East Germany and West Germany—into which the nation was divided after World War II—were reunited in 1990. Prussia, Silesia, and parts of Pomerania, which were German territory until 1945, are now part of Poland. Some of the other German regions and important cities are also shown.

The wealthy German at left employed the other two men to cook and care for his garden. Such working-class Germans often went to America in search of better opportunities and land of their own.

HOMELAND

Francis Daniel Pastorius, leader of the first organized German settlement in the United States, disapproved of conditions in Germany in the late 17th century. He called the life of the German upper classes the "Great Idleness."

Many thousand Germans, mostly of the nobility... are accustomed to follow the vanities of dress, speech, foreign manners and ceremonies, and incur incredible expense in learning to mount, to ride, to dance, to fence...while not a single thought is given to the love of God and learning to follow Christ.

John L. Betz, born in 1792 in the village of Flieden in the state of Hesse, came to the United States in 1833. He kept a journal that began with his life in the Old Country. In it Betz described the hard times that Germans faced in the early 19th century.

After a long devastating war that lasted almost a quarter of a century, the all-merciful providence gave us the golden peace. But even in the enjoyment of peace people were often burdened with common evils. [In 1815 and 1816] frequent rains caused not only heavy flooding but also damaged the fertility of the earth and reduced the produce from it which nourishes the people.

A rise in costs occurred in the years 1816 and 1817 that assumed unusual proportions and caused sad consequences which prevailed all around over a large area so that the grain required for the essentials of life had to be gotten from far-away Russia....

This was a punishment from above, I think.

In 1836, Henriette Geisberg Bruns came to the United States with her husband to settle in Missouri. Born in 1813, she spent much of her young life in the town of Oelde, Westphalia. In her autobiography she described her early life in Germany.

Now I begin to remember how I was often at Grandmother's and how she gave me presents and loved me. A Christmas tree, which in those days was very rare, glittered with many candles at Christmas time. At her death in 1820 I felt my first great pain.

From then on I remember how much the stork stopped in at our house. We were seven brothers and sisters. Franz, a weak child, Heinrich, Bernhard, Johanna, Wilhelm, and Therese. We children had a happy youth. Father was serious. He had a considerable library and read a lot. Mother was lively, energetic, and the soul of the house. She was always busy in the household with the servants and with us children and in the garden; she was ambitious and inventive. I remember that I attempted to help her in the garden. As a rarity she had several kinds of Indian corn as well as some colorful beets....

We had a piano, a grand piano from Grandmother, who was probably a virtuoso. There was a box full of music beside it. At an early age I received lessons in piano and particularly in voice. Mother and I sang duets, I always alto, and since she was a member of a singing society, I helped her study her part. That made me proud.

In school I learned well, I believe, and I was also instructed at home. In class I was always the youngest but not always the best-behaved, since I remember being punished several times. As a hobby I cut quills for writing, and for that purpose I always had a fine sharp knife which earned me the thanks of my classmates. And when Father once asked me for a sharpened quill, since he no longer could see so well, I felt very important.

In school...we also learned to sing from notes, and I did well in that and was allowed to sit next to the music teacher in church to assist her; she had asthma. The boys sat with a teacher in the choir loft, and we sat below. We took our turns singing. I loved the songs from the Verspoell song book very much. I always remembered them, and even now I often hum these melodies.

German workers in a pottery shop. Such skilled craftspeople found a demand for their talents in the United States.

Carl Schurz was born in 1829 in the village of Liblar on the left bank of the Rhine River. After participating in the Revolution of 1848, he came to the United States, where he served in the Civil War and later as Secretary of the Interior. He described his home village in his autobiography.

I must have been a little over four years old when my parents left the castle to establish a home of their own in the village of Liblar. The village consisted of one street. Midway on an elevation stood the parish church with its pointed steeple and cross. The houses, mostly one-storied and very small, were of whitewashed plaster, with frames and beams exposed, and tiled roofs. There were perhaps half a dozen brick buildings in the village, belonging to the count. The inhabitants of Liblar, small farmers, laborers, mechanics and a few inn- or shop-keepers, took an especial pride in their village because its street was paved with cobblestones. Notwithstanding our house had two stories, it was very small, with ceilings so low in the upper story that my grandfather when standing upright almost touched them with his head....

I was ten years old when my father took me to the gymnasium [a school that prepared young men and boys for a university education] at Cologne.... In those days, Cologne had about ninety thousand inhabitants, and was, as I supposed, one of the finest cities in the world. My grandfather had taken me there several years before on a visit, and well do I remember

Street fighting during the 1848 revolution in Berlin. The failure of this and other political revolts caused many Germans to leave for the United States in the 19th century.

the two things that then interested me most: the cathedral tower with the huge crane on top, and the convict chain-gangs sweeping the streets—sinister-visaged fellows in clothes striped dark grey and yellow, with heavy iron chains on their feet that rattled and clanked dismally on the pavement stones, one or more soldiers standing guard close by, gun in hand. I remember also how my grandfather reproved me for taking off my cap to everybody whom we met in the streets, as was the custom in our little village at home; for he said there were so many people in Cologne that were one to attempt to bow to them all there would be no time left for anything else; that one could never become acquainted with all those persons, and many of them were not worth knowing; and finally, that such deference on my part would mark me at once as a country boy and make me appear ridiculous.

Oscar Ameringer was born in the town of Achstetten by the Danube River in 1870. In his autobiography he described his time as an apprentice.

I became an apprentice in father's cabinet shop. The machine age had already come in, but as far as our shop was concerned we were still in the handicraft era. Furniture was not made wholesale, but by order.... The customer selected what he wanted from drawings made by hand. Price was set by the quality of wood and workmanship. Slighting workmanship for speed was a deadly sin. The customer's selection made, we transferred the drawing to wood, sawed out the wood by hand, dressed it, put it together, planed, sheaved, sandpapered it until the last blemish had vanished, then polished it by hand and finally delivered it by wheelbarrow or pushcart, propelled by hand.

It was slow work. I remember it took us a week to make an ordinary dresser, and two weeks or better for a fine or extra-fine one....

Jewish woodworkers outside a Berlin workshop in 1889. Many German immigrants found that their skills at trades such as carpentry and shoemaking helped them prosper in the United States.

Father was a guild master. Indeed, he was more than that. He was the master of the guild of joiners of the town. At the height of his career, father employed four journeymen and two apprentices. Master, journeymen and apprentices shared the same roof and ate at the same table. The best went to the master. Second best to the journeymen. Apprentices took what was left, paid for their tuition, and for good measure, washed dishes, ran errands, watched babies and changed diapers.

When an apprentice had served his three years, he made his "journeyman piece" in the shop of another master. If then, in the opinion of a committee of masters, he had demonstrated the proper qualifications, he was sent on the *Wanderschaft* [journey] to perfect himself in his craft in other cities and lands. The wandering journeyman was not a hobo. He was entitled to all the rights and privileges of a journeyman and future guild master. Entering a workshop of his craft he would say, "God greet thee, masters and journeymen of my guild." That done, and no work available, he would receive a definite, stipulated journeyman's stipend from his fellows, double the amount from the master, and so continue his journey.

In the larger towns there were *Herbergen* [shelter homes], houses in which guild journeymen found welcome overnight. In the larger cities there were shelters for each separate guild. When the journeyman returned and passed his mastership, he was crowned master.

A young woman identified as Agnes M. gave her story to the American magazine Independent *in 1903.*

I was born just twenty years ago in the old, old city of Treves.... There were eight children in our family, five girls and three boys....

My father was a truckman, carrying goods from the railway stations to the shops; he had a number of wagons going and had built up a good business, though he was always ill from some disease that he contracted when a soldier in the war with France. It was consumption [tuberculosis], I believe, and it finally carried him off [when Agnes was three]....

By the time I was five years old my mother had lost everything except the money she got from the government, which was enough to keep her, but the family had to break up, and I went away to a school kept by Sisters of Christian Liebe [Love], in another city. The government paid for me there on account of my being a soldier's orphan—all of us children had allowances like that.

From the time I went away to that school till I was fifteen years of age I did not once see my mother, but stayed in school during all the holidays....

We got up at half-past six o'clock each morning, and had mass three times a week and morning prayer when there was no mass. At eight o'clock school began and lasted to ten, when there was half an hour for play, then an hour more

German coach makers in the 1850s. This was a trade at which German Americans excelled. Some, like the Studebaker family, later expanded into automobile manufacturing.

German papermakers in the 1850s. The first paper mill in the United States was built by William Rittenhouse, a German immigrant, in 1690 in Roxboro, Pennsylvania.

Otto von Bismarck

When the Northern Pacific Railroad built tracks across the sparsely populated Dakota Territory in 1873, a land company ambitiously laid out a new town in what is today North Dakota. Hoping to attract German immigrants to settle and populate the town, the company named it Bismarck. It is the only U.S. state capital named for a foreign ruler.

Otto von Bismarck was then the hero of the German people. He had achieved what Germans had wanted for hundreds of years—unification of the nation under one ruler. As prime minister of Prussia, the largest German state, Bismarck had made allies of the other states in a war against France. After the Franco-Prussian War of 1870 ended in a German victory, he persuaded the other German rulers to accept the Prussian king as emperor of a united Germany.

In fact, William I, the first German emperor, was only a puppet controlled by Bismarck. For the next two decades, Bismarck set the foreign and domestic policies for Germany. Some of his plans were benevolent. He instituted, for example, the first comprehensive social security program in Europe, giving workers insurance against accidents, illness, and old age.

On the other hand, Bismarck could be ruthless in protecting his power. When a Roman Catholic political party gained a large number of seats in the Reichstag, or parliament, Bismarck felt threatened. He launched a *Kulturkampf*, or "culture struggle," to protect Germany from what he saw as an external power. Catholic schools were closed, and some bishops and priests imprisoned.

German Catholics responded to the *Kulturkampf* by emigrating in sizable numbers during the 1870s and afterward. Many came to the United States, settling in such midwestern cities as Milwaukee, Cincinnati, and St. Louis.

In 1888, William I died and his grandson, William II, became emperor. The young ruler and his aged prime minister soon clashed. After Bismarck attempted to create a crisis that would allow him to seize power, William forced him to resign. Yet Bismarck's legacy lived on in a united Germany that was one of the great powers of Europe.

school, then more play and then lunch, after which we worked in the garden or sewed or sang or played till six o'clock, when we had dinner, and we all went to bed at eight. We did not always go to sleep though, but sometimes lit candles after the Sisters had gone away and had feasts of apples and cakes and candies.

There were about eighty boys in this school and fifty-five girls—none of them older than fifteen years. We had a very large playground, and though the boys and girls were kept separate they yet found means of conversing, and when I was eleven years of age I fell in love with a tall, slim, thoughtful, dark-haired boy named Fritz, whose parents lived in Frankfurt. We used to talk to each other through the bars of the fence which divided our playground. He was a year and a half older than I, and I thought him a man. The only time I was ever beaten at that school was on his account. We had been talking together on the playground; I did not heed the bell and was late getting in, and when the Sister asked what kept me I did not answer. She insisted on knowing, and Fritz and I looked at each other. The Sister caught us laughing.

Whipping on the hands with a rod was the punishment that they had there for very naughty children, and that is what I got. It did not hurt much, and that night at half past nine o'clock, when all the house was still, there came a tapping at our dormitory window, and when it was opened we found Fritz there crying about the way I had been whipped. He had climbed up one of the veranda posts and had an orange for me. The other girls never told. They said it was so fine and romantic.

Walter Hoops was born in Hanover, Germany, in 1902 and came to the United States in 1927. In 1972 he told an interviewer about his early life.

When I was a young man in Germany, I belonged to the Wandervogel movement, [which means] the hiking birds...young people [who] threw their stiff collars away. We were the hippies of our age. We also had long hair. It always seems to have something to do with hair. We also got away from all the stiffness of clothing, and we used to wear short pants and comfortable shoes when we hiked, and we were just looked down on and made fun of.... But still we were different. I remember the first time I didn't even comb my hair properly...[without a part] but just combed back.... [It] was practically a revolution to my father and mother.

John Otto Reinemann fled to the United States around 1933 when Hitler began his persecution of the Jews. In old age, Reinemann recalled the sorrows that World War I had brought to Germany. At the beginning of the war, he was a student in a gymnasium in Frankfurt.

Our school was part and parcel of the Prussian educational system and, especially during the war years, it was not to be outdone in patriotism and nationalism. A number of our younger teachers had been drafted or had volunteered for military service, and the longer the war lasted, the closer we students came to the age at which we would be recruited into the Army. The defeat of Imperial Germany in 1918 saved us, and I owe it to my lucky star that I was too young in the First, and too old in the Second, World War to have ever become a soldier....

At the beginning of the war, a soldier was quartered in our apartment. A peasant from the nearby Rhön Mountains, he received his military training in Frankfurt. After a few weeks, he was sent to the Western Front and we mailed packages of food and other necessities to him. We regularly received his laboriously written postcards. He had adopted us as his "next of kin." He visited us on his furloughs, until one day our mail to him was returned with the annotation: "Private Kirchner killed in action." We children had come to like him for his kind and simple ways, and the news of his death caused us great sadness.

In the later years of the struggle, when the armies were immobilized in the stalemate of trench warfare...the German civilian population increasingly felt the brunt of the war. Food and clothing became scarce, and fuel, electricity, and all other commodities were rationed. The cold months of 1916 and 1917 became known as the "Turnip Winters," because this was the only agricultural product still available with which to make soup or cook a vegetable dish.

Walter Lowen left Germany in 1936. Nearly half a century later, while living in Seattle, he recalled the decision that led him and other Jews to escape the Nazi regime.

I was born in Munich and lived in Munich for thirty-three years. My family goes back in southern Germany since 1740, which is quite a few years....

I could see the danger, being so close to Hitler and the Nazis in Munich. [Hitler had built the Nazi party and organized his first attempt to seize power in Munich in the 1920s.] Many of my school friends, former school friends, were becoming big Nazis....

Naturally I had to give everything I had, except my furniture, which they allowed me to take with me. But I was not afraid. I had a good education, and I still believe it's what you have in your head, it's what helps you wherever you go.

Before I left, one of the unforgettable evenings was the last concert Bruno Walter, at that time the head of the Munich Symphony, gave before he had to leave Germany. They played Beethoven's Ninth Symphony, and the goodbye applause lasted over an hour in an audience mostly of gentiles. He was one of the first ones who had to leave. There was nothing one could do.

In the 1930s, many German Jews, fearful of Hitler's anti-Semitic policies, tried to obtain visas to enter the United States. Tragically, the U.S. government refused to raise the quota on immigrants, and many were turned away. Eric Cornell described his experience:

I started procedures with a visit to the American Consulate in Berlin. The place was jammed with applicants for a visa, obviously scarce as hen's teeth....

I finally came face to face with the consul, eager to show my academic knowledge of the English language. Will you believe it? He asked me, 'What is hollowware?' [It refers to a kind of table utensil.] I do not recall what silly answer I gave him, but I was then and am now convinced that such a silly question was designed solely for turning me down, which on account of the ridiculously low quota he would have done anyway.

Cornell later made his way to the Netherlands, where he found a more sympathetic consul who gave him the necessary visa.

The Silberstein family, around 1880, when they left Germany for the United States. They later settled in Iowa.

GERMANS FROM RUSSIA

A German woman threshes a crop of hemp in Volhynia, in today's Ukraine. West of the city of Kiev, Volhynia was one of the regions of the Russian Empire where Germans settled. In the late 1800s, many Germans left Volhynia for the United States.

Catherine the Great became empress of Russia in 1762. She encouraged immigrants from Germany to settle along the Volga River and near the Black Sea. Johannes Fischer moved from Württemberg to southern Russia in 1832. His grandson, Christian, was born there and later came to the United States. Christian recorded his memories of that period.

The Fischer family moved to Russia because of the promise of free land: sixty *desyatina* of good farm land had been promised to each head of the family. Since one *desyatina* was 2.7 acres, this meant over 160 acres of land [the amount offered to Americans in the Homestead Act of 1862] for each family. Commissions had been set up by the Russian government to help the German settlers, and the immigrants received what had been promised them. The land proved to be very good for farming, and they had no problem borrowing money at the bank in Odessa. They were able to add thousands of acres to their holdings by buying from the Russian nobility who owned vast tracts of land.

The German communities in Russia were organized by religion. Sarah Harder Warkentin was born in a Mennonite settlement in 1858. As a young girl, she took care of the younger children. She described the death of her favorite brother, Peter.

We got another little brother, Peter. He was such a sweet baby, but when he was eighteen months old, he began to get sickly. He could talk already. I took care of him all the time, and when I wasn't with him, he would ask so pitifully, "Where is my Sarah?" He didn't seem to care about anyone but Mother and me. We were his best friends in this world.

Then in the fall and winter, I had to go back to school. When I got home, I had to keep the stove going. The houses were heated with large brick ovens that burned straw. It took about two hours, then it would keep the house warm for half a day or longer. I was with Peter as often and as much as I could be....

That summer I got sick. I had chills and fever. Little Peter kept on getting worse until he finally had to stay in bed. One day Mother had put him in front in a big bed, and I had crawled in back of him with my fever and had gone to sleep. Then all at once I heard Mother say, "Father, Peter is dying!"

I jumped up and out at the foot of the bed and went into another room and cried as if my heart would break. Then a neighbor, a Mrs. Fast, came and took me out at the other end

Alexanderfeld was a small German farming village in the Black Sea region of the Russian Empire.

of the house and said, "You won't even let the poor child die." So I didn't even see him die.

It was customary for the men who worked in the fields to rest for an hour at noon. They worked from daylight till dark. It was just at the noon hour and Father and my brother Martin were home resting. This was on a Friday, July 20, 1868.... So Father and Martin could not go to the fields anymore that day. They had to dig the grave. All the men were working in the harvest.

Mr. Jenzen always had a cabinetmaker in his shop, so he made the coffin. We had to pay for that, but Mrs. Jenzen sent word that she would send the zwieback [a crisp bread] and ground coffee for the funeral. [These were the customary refreshments served after funerals.] Mrs. Jenzen also came Sunday morning and dressed little Peter. She dressed him so nicely in pants and a coat. He had never worn pants yet. Sunday was the funeral. As they didn't work in the fields on Sunday, all the men living in the village could come to the funeral.

When the Germans had been invited into Russia in the 18th century, the Russian government promised that they would be exempt from military service. But the tsar repudiated that agreement in the 1870s. Some Germans started to leave—particularly Mennonites, whose religion forbade the use of violence. Many of the young men who remained were drafted into the army. Ludwig Neher, who came to the United States in 1909, told his daughter about his experiences during the Russo-Japanese War of 1904–05. Pauline Neher Diede related his story later.

German-Russian children in the Black Sea colony of Neudorf. There were about 200 German villages in this part of the Russian Empire by 1850. Typically, each was composed of people of one faith—Lutheran, Roman Catholic, or Mennonite.

Cossacks [Russian cavalry] had been riding through village streets picking up young men without serving them notice. Mothers cried and screamed at them. Ludwig hardly had time to say good-bye to his mother and gather up a few belongings. His mother shoved into his bundle

German Russian workers stop for lunch during the grain harvest. Though the soil in the Black Sea area was fertile, rainfall was scanty and crops sometimes failed.

the warmest pair of underwear, and Ludwig often said later that if it had not been for those underwear, he would have frozen to death. Military training was not very rigorous. Ludwig remembered doing things as clumsily as he could, purposely bungling everything. He was assigned to the kitchens, and found that cattle in feedlots were treated better than soldiers. One day his job was to get the sauerkraut out of the barrel. It was a hot summer and worms were crawling through the kraut, but when he described the mess to the cook, he got a blow on his head and orders to cook the worms with the kraut.... For the rest of his life my father hated even the smell of sauerkraut.

He served about a year, through a record cold winter and an especially sweltering summer. He saw two buddies killed, and another severely wounded, whom he carried to a pond. Here he washed his friend's wounds and fetched him drinking water in his cupped hands. Ludwig often spoke of that man's miraculous recovery, for after two days and nights in their hideout, a peasant woman appeared, as out of thin air, dressed in rags, bringing a kettle of cooked grain and a spoon. She came from a village nearby, and must have heard their moans.

Sister Reinhardt Hecker was born in 1901 in München, Russia, a village near Odessa. She came to the United States with her family in 1914. In a 1992 interview she described the Christmas festivities of her girlhood in Russia.

That was a great feast, nobody worked, not even my folks. No washing was done until after the sixth of January because then the feasts were over. But it was one party after the other, they [went] from place to place [to] have a party. We had a great big room.... My dad played the accordion. And they had a few dances, just friends that came, you didn't invite anybody, they came by themselves....

Germans living in the Volga River region of the Russian Empire thresh barley by striking the sheaves against a wooden table. Threshing separated the seeds, from which flour was made, and the stalks.

A wedding celebration in the German colony of Arzis, Bessarabia. The bride and groom stand arm in arm at the center. Catholic and Lutheran weddings were festive occasions, often lasting for three days of dancing and feasting.

They did a lot of baking. I still remember the month they made *Zurker Kuchen* [Christmas cookies]...just stacks [of them] in the oven there and we had a big trunk lined with tin and there was stored...the fancy bread and all that *kuchen*. And I had to go up and get it sometimes, once in a while I had to count how much there's left. By the time those feasts were over everything was gone because they used them for parties and everything....

At Christmas Eve, a girl was dressed in white, with all kinds of lace over her head with a gold crown, and she would come into the house and ask if the children were good. My dad said, "yes, come in." (The girl was the *Krist Kindle* [Christ child].) They made us pray, then gave us a basket which consisted of toys, candy, and nuts....

When I was older I could go along [to Mass on Christmas Eve]. And when they came home there was a ham on the table and a big knife and everybody got to start eating again. And at Christmas night they had a brass band playing in church and the balcony and singing. They had lanterns hanging all around. Our church had a big wall around it and the parish house and they had all those lanterns because there was no electricity. And of course...all the German singing.... And the next day they had high mass again, just like they did it in this country.

German emigrants who traveled to port cities stayed in reception halls such as this one in Emden until it was time for their ship to leave for America.

CHAPTER TWO

GOING TO AMERICA

Few of the German states were seafaring powers, and as a result the colonization of America was carried out by other nations. However, some Germans played a role in it. There were, for instance, Germans among the English colonists at Jamestown, Virginia, in 1607. Prussian-born Peter Minuit (whose name in German was Minnewit) became the first governor of the Dutch colony of New Amsterdam in 1626. It was Minuit who purchased the island of Manhattan—today perhaps the most valuable piece of real estate in the world—from local Native Americans for trade goods that were worth about 60 Dutch guilders (the equivalent of $24).

The first large group of German immigrants came from the Rhineland, the area that had suffered most during the Thirty Years' War of 1618–48. On October 6, 1683, 13 families from the town of Krefeld arrived in Philadelphia on the ship *Concord*. They had been invited by William Penn, an English member of the Society of Friends, or Quakers, who had founded the colony of Pennsylvania a year earlier. Wishing to populate this vast tract of land with European settlers, Penn visited the German states to encourage emigration, offering religious freedom and farmland.

Most of the passengers on the *Concord* were Mennonites, a Protestant sect whose practices and beliefs were similar to the Quakers'. Having endured religious warfare in Europe, the Mennonites were pacifists who opposed all forms of violence. Their leader, Francis Daniel Pastorius, had arrived earlier, declaring his intention "to lead a quiet, godly, and honest life in a howling wilderness." These first German Americans established a community called Germantown, which still exists within the boundaries of Philadelphia.

Many more German peasants followed during the 18th century. Those who had no money for their trans-Atlantic passage arrived in America as "redemptioners," or indentured servants. They agreed to work for a period of four to seven years to pay off the cost of their ship passage. American colonial landowners came aboard the newly arrived ships to purchase redemptioners in a system that was like temporary slavery. Indeed, colonial newspapers were filled with advertisements offering rewards for redemptioners who had run away from their masters.

The journey to America in colonial times was uncomfortable at best and deadly at worst. One German who crossed the Atlantic in 1728 wrote in his diary that the food on ship "consisted of horrible salted corned pork, peas, barley, groats, and codfish. The drink was a stinking water, in which all food was cooked."

The time it took to cross the Atlantic varied greatly, depending on the time of year and the weather. Gottlieb Mittelberger, who emigrated in 1750, wrote that one ship took six months to cross the stormy ocean in winter. Of the 340 persons who had sailed in it, only 21 survived the voyage. Mittelberger noted that many ships sank in midocean, a fact he claimed was concealed so that future emigrants would not be discouraged.

The agents of shipping companies and recruiters for the American colonies made extensive efforts to attract immigrants. They traveled through the Rhineland in brightly colored wagons. Drawing a crowd with trumpets and drums, the recruiters described in glowing terms the life that awaited in America.

In addition to the dream of free farmland, Germans came in search of religious tolerance. Besides the Mennonites, many Lutherans and Reformed Church members also arrived in colonial America, often coming from the German states Bavaria and Würzburg, where

Catholicism was predominant. Smaller numbers of German Catholics also arrived, such as a group expelled in 1732 by the staunch Protestant Count Leopold of Firmian.

Estimates of the total number of Germans who arrived in America in colonial times range from 65,000 to 100,000. The final group were deserters from the German forces who fought for the British in the Revolutionary War. Their story illustrates just how miserable a life the German peasants of that time led. Most of them had been forcibly pressed into service by the princes of the domains in which they lived. The princes then sold their subjects' services to the British for large cash payments. Because about half of these unwilling recruits came from the state of Hesse-Kassel, all came to be referred to as Hessians. During the Revolution, some of them deserted and joined Washington's army. At least 5,000 more Hessians greeted the news of the American victory by deciding to remain in the new country as citizens.

The pace of German emigration slowed from 1790 to 1815, in part because of the Napoleonic Wars that then engulfed all of Europe. With Napoleon's defeat, Germans in rural areas again began to seek relief from poverty, famine, and overpopulation by emigrating.

Religious persecution once more played a role in the push to emigrate. From about 1830 to the 1880s, anti-Semitic laws were passed in some German states. During that time, many German Jews left to seek a better life in the United States. Germans were the largest group in American immigrant history to include sizable numbers of all three major Western religions: Catholicism, Protestantism, and Judaism.

In the early 19th century, the majority of German emigrants were young, unmarried males. Later in the century, entire families made the journey together. A French writer in the 1840s described the "lamentable sight" of Bavarian villagers traveling toward the French port of Le Havre: "The

This monument in Bremerhaven, Germany, commemorates the German emigrants who departed for America. After World War II, refugees to the United States from the former German territories in the east contributed funds to pay for the statue.

long files of carts that you meet every mile, carrying the whole property of the poor wretches, who are about to cross the Atlantic...piled with the scanty boxes containing their few effects, and on the top of all, the women and children, the sick and bedridden, and all who are too exhausted with the journey to walk."

Guidebooks for emigrants and letters from those who had gone before continued to spread word of the vast fertile lands and free society that awaited them in America. Family members who prospered in the United States sometimes sent home the money to bring others across the ocean. By now there were numerous German American communities where immigrants could find familiar customs and others who spoke their language.

The second wave of German immigration peaked in the decade 1850–59, when nearly a million Germans arrived in the United States. During each of the next two decades, about 700,000 German immigrants landed. Then from 1880 to 1889 another peak was reached when more than 1.4 million German speakers arrived. Sizable numbers in this latter group were Catholics seeking refuge from Bismarck's *Kulturkampf* described in Chapter 1; others were Germans from Russia who fled that nation when its government adopted a repressive policy toward them.

Well-educated Germans, some of whom had attended universities, fled during various periods (1819, the 1830s, the late 1840s) when the German state governments sought to repress liberal movements that attracted students, teachers, and intellectuals. The "'48ers"—those who had taken part in the revolutions of 1848—made notable contributions to their adopted country.

The process of emigration could be difficult. The German states required all family members to obtain emigration visas. Emigrants had to provide such documents as baptismal and marriage certificates from their local church, evidence of having a trade or profession, and proof that adult males had fulfilled their military service. Charitable organizations, in cooperation with government agencies, sometimes supplied emigrants with a travel guide, a map of the railroad lines in the United States, and a list of German settlements where they would be welcomed.

Until about 1830, most emigrants sailed from the Dutch port of Rotterdam or the French city of Le Havre. Afterward, German ports became the major departure points. Steamships, and later railroads, carried people to the port of Bremerhaven, in Bremen. Most of the emigrants from southern and western Germany departed from here. Indeed, by the mid-19th century Bremen had become known as *Der Vorort New-Yorks*, "the suburb of New York."

A few decades later, Hamburg, on the Baltic Sea, attracted emigrants from eastern and southern Germany. During the 20th century, Hamburg became the chief port of emigration as large passenger-carrying steamships established regular routes across the Atlantic from there.

In every port from which sizable numbers of emigrants left, inexperienced travelers faced swindlers who tried to strip them of their savings. In 1870 the governor of Minnesota, which had a large German immigrant population, sent Albert Wolff to investigate conditions in Bremerhaven. Wolff noted that there were about 70 emigrant boardinghouses in the area. But he said the emigrant was "preyed upon by blacklegs and confidence men, here called *Bauernfanger,* which means trappers of 'green horns,' from the moment he starts from his native city or village."

Officials were not immune to bribery. For example, when Edward Stein left for America in 1871, to avoid serving in the army of Prussia, he boarded a ship in Hamburg without a passport,

The Gros family gathered in Hellenhahn, Westerwald, Germany, in 1900 for a portrait. Clemens, standing fourth from left, was about to leave for Texas.

which he could not obtain without proving he had served in the army. When a ship's officer asked for his papers, Stein handed him a packet of 400 marks of German money that his father had given him. "I find the papers of Mr. Stein to be quite correct," the officer declared. (Later, Stein would become the first mayor of Pocatello, Idaho.)

A majority of emigrants traveled in the cheapest berths, the steerage. There, poor ventilation, over-crowding, unsanitary water, and a minimum of food encouraged the spread of diseases such as cholera and scurvy. In 1853, a New York newspaper called the immigrant vessels "plague ships and swimming coffins." In an attempt to foster good health, German ship captains required emigrants aboard to take exercise on deck.

As the age of steamships dawned, the time of a trans-Atlantic trip was cut from six weeks to two or three weeks, but few emigrants could afford the more expensive steamship passage until late in the 19th century. German immigration continued to exceed 300,000 in every decade until 1930, except between 1910 and 1919, when World War I temporarily halted it. Even so, from 1923 to 1963 the number of German arrivals on the shores of the United States still outnumbered those from any other country. During the 1930s, many were Jews seeking refuge from Hitler's anti-Semitic policies.

After 1945, another wave of German emigrants came flooding into the United States to escape the devastation caused by World War II. The postwar division of Germany into East and West Germany brought about half a million additional refugees, most fleeing Communist East Germany.

In the three centuries since the first German American settlement was founded, about 7 million German-speaking emigrants have made their way to America. They and their descendants form the largest single ethnic group in the U.S. population today.

A 19th-century engraving titled The Emigrant's Letter shows a German farmer stopping work to hear his wife read a letter from a friend or relative in the United States.

The cover of a guidebook to Wisconsin, distributed to prospective immigrants in Germany, was published by the State Board of Immigration of Wisconsin. Like other sparsely populated midwestern states, Wisconsin wanted to encourage immigrants to settle.

THE DECISION TO LEAVE

In 1843, the Bohning family left Barkhausen, a small village in northwest Germany. In 1890, Ernst Bohning, who had been 10 at the time, recalled his family's decision to leave.

Whenever we asked what America looked like and what the living conditions were like, we got contrary comments. The usual reply was, "So, you're thinking of going to America? I wouldn't have thought you'd consider that. My advice is stay in this country and earn your food honestly. You know what you have here. The main question is, are there enough boiled potatoes? Get this stupid idea out of your head. That's my advice."

We knew very well what we had, namely six acres of good land, located by the river, and good enough to get by on. Nevertheless my Father was not satisfied, for life went on in an endless cycle of manure wagon, spinning wheel, pumpernickel, sour milk and boiled potatoes. What's more, there were nine children in the family, five of them boys. If we stayed, they might be drafted in the next war, and be shot to death for the King and the Fatherland. Father had been a soldier himself, and he did not want that fate for his sons. So the thought of leaving remained in his head....

It so happened that Ernst Borges [a neighbor]...decided to emigrate with his young family. Father declared that his oldest son, Heinrich, who would soon be drafted, would join Borges.... This was the advance scout of our family, in a way, a reconnaissance patrol. We could rely on what they would report, and that would determine what the whole family would do....

We anxiously waited for the first reports to arrive. In August they came, and were favorable. They convinced my father.... At last, we kids shouted, "Hooray, we are going to America!" We had no idea what lay ahead of us on the long journey.

Emma Murck Altgelt went to Texas from Germany in the middle of the 19th century. She remembered being influenced by the stories told by a relative who had returned to Germany for a visit.

He told many enthusiastic stories about his new home. I was thrilled when I listened to him describing the charms of that recently opened paradise, the eternally blue sky, the radiant sun about the great uncultivated, uninhabited land, tempting tropical fruits, Indians and wild animals, too, to break the monotony of existence and above all else, golden freedom. Texas became the land of my dreams.

Richard Bartholdt, who became a U.S. Representative from Missouri, was born in Germany in 1855. The ideals of freedom were what propelled him to come to the United States.

My father was a "Forty-eighter," one of the German patriots who strove for democracy and a united country.... Under his tutelage I became imbued with democratic ideals and an intense love of liberty even as a boy. Hence, I may fairly claim to have been an American in spirit even before I set foot on the hospitable shores of this country. He hated militarism and said he did not want his boy to become "cannon fodder." This fidelity to principle wrung from his unwilling heart the decision to let me go to the new world, while, owing to the death of my good mother and sisters, he was left alone in the old.

A young woman who gave her name as Agnes M. described to the Independent *magazine why she left Germany at the end of the 19th century.*

I heard about how easy it was to make money in America and became very anxious to go there, and very tired of making hats and dresses for nothing for a woman who was selling them at high prices. I was restless in my home also....

At last mother, too, grew anxious to see me go. She met me one night walking in the street with a young man, and said to me afterward:

"It is better that you go."

There was nothing at all in my walking with that young man, but she thought there was and asked my eldest sister to lend me the money to go to America...and a month later I sailed from Antwerp, the fare costing $55.

George Grosz, born in Berlin in 1893, became famous in the 1920s for his satirical drawings. He came to New York in 1931 to escape the oppression of the Nazi government in his native land. He recalled hearing descriptions of America by other immigrants.

Sometimes people returned from the United States, usually for just a short trip. Such visits always created a miniature sensation, since the visitors always gave parties and spent the bit of money they had saved and brought with them in the form of travellers' checks. We would admire the brand-new suitcases with their many ship tags, the new suits, patent-leather shoes, pipes and "half-diamond" cuff links. I can still see Jim, whose name in Germany had been Fritz, pulling some coins out of his pocket and letting them roll on the table as he carelessly said, with pipe in mouth, "Los! lass ma anfahren" ("All hands to the bar"). Long after such a miraculous "uncle" had disappeared we continued to talk about him and wished to be like him.

In the years following World War II, U.S. social agencies sent packages of food and other gifts to needy people in Europe. One of these caused young Sabine Reichel in Hamburg to fall in love with America and emigrate there in 1976.

How did it all start?... I still think it was the pink rabbit slippers that made me fall in love with America.

"Pink rabbit slippers?" comes the puzzled reply, and I add, still with a sense of pride, "I grew up with packages from America." Those heavy brown boxes from a dreamland far, far away, often torn from a bumpy trip on a freighter, have shaped my childhood as much as did the ruins and cripples of my native land. For me, those boxes were a symbol of freedom and superiority....

But more than just material treasures, they reached out through the spirit of giving and made me feel special. A country that was inhabited by such generous people who sent wonderful presents to two little German girls they had never met had to be paradise. I was overwhelmed with gratitude and affection and became hopelessly infatuated with America.

Henry Baum built this fine house in Frank, Russia, after his return from the United States. The prosperity of those who returned home prompted others to try to emigrate.

LEAVING HOME

Theodor Engelmann was born in the Palatinate in 1808. After taking part in a student uprising, he was forced to leave Germany in 1833. He described the warm farewell other students gave to him and to fellow revolutionary Gustav Koerner.

Flags decorated the well lighted hall, toasts were given, glasses clicked, fiery speeches were made until it was time to take up our reserved seats in the post [mail] coach to Paris. Now they gave us our passes. Koerner's pass was in the name of "Chretien Huescheler," commercial traveler, if I remember correctly. Mine was in the name of "Victor Klenck," notary clerk. After short, hearty farewells from our kind hosts and accompanied by their good wishes, we climbed into the post and rolled into the dark night.

Henriette Geisberg Bruns left Germany in 1836 for Missouri, as she described in her autobiography.

All preparations were made.... Everything that had to be taken along was packed in boxes, and the other household things were sold.... Bruns [her husband] was getting his key from the post office across the street when he received a letter. After we had light, he opened the letter in the empty house. It contained the news of his mother's death. That's all we needed to shake our forced composure. We had a very bad night. After that the last things were done quickly. Our brothers arrived, and travel companions reported in and then traveled ahead. Our departure was set for Sunday morning.... The brothers found the wagon the eve before. When it arrived they jumped in, and we followed and away it went. Amazed and perturbed, the neighbors looked after us when they realized that we were waving good-bye. Our little son, who had just gotten rid of the measles, got very sore eyes upon the sandy roads. To be sure we had taken the road to Bremen by way of Bruns' home in order to bid farewell to the relatives on our way. At the Eylardi Inn in Bremen, we waited for the arrival of our goods, which were loaded on heavy wagons and traveled at a slower rate.

Here, a young man bound for the United States in the 1850s bids farewell to his family and friends.

In 1874, Pieter R. Voth came to the United States as a child from the village of Gnadenthal in southern Russia. The family left with the rest of their Mennonite congregation to avoid military service, which was opposed to their religion. Voth described the departure.

So after selling the house and the land, many articles which were supposed to be needed in the new world, were packed into large boxes, trunks and bags, to take along overseas. I can remember how the running gear of our farm wagon was taken apart and fitted into the smallest box possible with smaller articles packed in between….

The day of the departure had come. A large crowd of neighbors and friends had come to say farewell. Someone had stepped on my toe so I had climbed into our neighbor's top buggy, in which our parents and smaller children were to be taken to the railroad station. I was crying, not because of the farewell, but because of the sore toe, and many people came and looked into the buggy to comfort me.

Naturally, father and mother took the farewell from home, country and friends quite hard. They did not know what the future might have in store for them and us. My youngest sister, Susanna, was only two weeks old when we left, so father had to carry her in a cradle-like box hung over his shoulder by a strap, half around the world. Our neighbor drove us to the railroad station while father and mother were quietly weeping. I am told that I said to my sister, Anna, who was six years old at the time, "Anna, do you know something; we go home now." This childish utterance, which I do not recall, is said to have greatly comforted our parents.

In the 19th century, travel on the rivers of Germany could be hazardous. Emil Mallinckrodt, born in Dortmund, Germany, in 1805, wrote a letter to his brothers and sisters describing a disaster on the Rhine during his trip to America in 1840.

Yesterday we suffered a shipwreck. At half past nine, a steamer from Dusseldorf rammed us and in eight minutes we sank eight feet underwater in the middle of the Rhine. It was dark and we screamed for help.

Soon boats arrived; so many jumped into the first two that I did not risk putting Ellen [his wife] into them. As the third came up, I carried Ellen into it and pushed off with mighty force that our boat would not be overfilled. We arrived safely and all were saved. Now one sees only the boiler and most of the ship. Two of our trunks were saved later on. That trunk from you, our traveling bag, and my trees lie at least ten feet under water. Will try to see to-day what can be done, but we were glad to escape with our lives. Ellen showed her usual courage and lies safely in bed in a country house….

On the second day, I had the big box and my trees fished out of the river so [we] lost nothing save the travelling bag with my new boots and many other pieces of clothing. Only yesterday, I was at the shore, to dry things partly.

A poster advertises the departure of a ship from Bremen, Germany, bound for Baltimore. Though the picture makes the ship appear sturdy, the trip was hazardous and some ships sank in the Atlantic.

A mother and her three children have their picture taken in Hamburg around 1914, just before leaving for the United States. Often, fathers left first, sending passage money for their families later.

The first leg of the Bohning family's trip to America in 1843 was over land to Bremerhaven. There, 10-year-old Ernst Bohning saw the sailing ship that would take them across the ocean.

We began the trip to Bremerhaven with our chests and boxes in a big hay wagon. We were joined by other farm families, including some young guys who were a lot of fun....

I had to sit on the edge of a chest with one leg stretched out and the other bent under me. What's more, we suffered from terrible heat and burning thirst in the linen covered wagon. The trip to America was already starting to look a little less bright to me....

Finally on the evening of the third day, we were there [Bremen]. Hungry, tired, stiff and sore, we stood at the corner of the restaurant where we were unhitched. Whimpering, we waited for dinner and bed.

Early the next morning, after we had eaten a really good breakfast, our belongings were loaded on a barge, which would go that afternoon to Bremerhaven. Meanwhile, we had time to see the town. We roamed around at random until we came to a beautiful, big church. Of all the sights we saw, we liked this best. We stayed there for a long time, for it was a Lutheran church, and we prayed hard for a good crossing. We also saw the big giant, Roland, at the town hall. "Roland the Giant" is a statue, standing straight and stiff by the town hall in Bremen. Somebody told us this phrase, and it stuck in our minds.

After we had lunch at our lodgings...we went on a barge that floated downstream on the Weser River. Oh, this was a

So many emigrants left from the port of Hamburg that a village was built to give them temporary housing.

lot better than being under the boxes and chests in the wagon. For awhile we were happy-go-lucky, then suddenly we got stuck on a sand bar. It seemed like the river was draining away. This frightened us children very much. But the sailors calmed us down and said, "In two hours there will be enough water, then we can go on." We hadn't heard about high and low tides before. They were right; in two hours, more water came into the Weser from the ocean, and we sailed happily on. It seemed wonderful to me that water could flow uphill.

The next afternoon we reached Bremerhaven, and our barge was brought to the side of a large three-master. At last this was the nice big ship that would bring us into the land of Canaan....

Fred Martin, who came to the United States from Russia in 1909 with his wife and children, described the train ride between Odessa and the port of Hamburg.

The train was crowded, as bad as a stock car of hogs. Children were perched atop sacks and bundles so closely they hardly had elbowroom, and they set up a howl. The palms of their hands were saucers into which was thrown a piece of bread and "schpeck" or salt-brined pork.... No toilet facilities except on the premises of a train station. Children rebelled and cried and were shoved, shouted at and slapped, even more by train conductors than by parents. I even felt like slapping my wife because she constantly scolded me for choosing this ordeal for the family.

Annemarie Ballin told an interviewer in 1983 about her difficulties escaping from Nazi Germany during the 1930s.

Some relatives in the United States gave an affidavit to my husband [enabling him to enter the United States].... My husband left Germany in March 1939, and took our daughter, ten years old, to England and left her there...with an English Jewish family....

When I finally got my visa, I had to take a train from Munich to Genoa, Italy.... One of the SS men [Nazi storm troopers] came through the train and said, "Everybody has to get out at the Austria-Italian border."...

The train got to the border. There were flowers, it was decorated, but the train didn't stop. In Milan, Italian soldiers came into the train. A soldier who could speak a little German said, "Hitler and Mussolini just met." That's why the train didn't stop, and why we didn't have to get off.

One misconception people have is that you couldn't leave Germany. You could always get out if you didn't take anything with you. You just couldn't get into other countries without visas. My family was killed—my mother, my sister, the whole family—in Auschwitz. I had no affidavits for them.

Albert Einstein

During the 1930s, thousands of Jews from Germany and other European countries fled from Adolf Hitler's anti-Semitic regime. Tragically, the United States refused to increase its immigration quotas to allow more of these refugees to enter the country, but exceptions were made for the most talented and educated.

Many scientists were among those permitted to enter the United States, and their contributions helped defeat Hitler's Germany in World War II. The most prominent among them was Albert Einstein, who had won the Nobel Prize in 1920 for his work in physics. He had published his theory of relativity in 1905, when he was employed in the German patent office as a "technical expert, third class." Few ordinary people could understand what Einstein's theory of relativity really meant, but he was popularly regarded as one of the world's most brilliant people.

Einstein moved to the United States in 1933, the year Hitler took power in Germany, and continued his work at the Institute for Advanced Study in Princeton, New Jersey. After Hitler invaded Poland in 1939, starting World War II, Einstein wrote a letter to President Franklin Roosevelt. Einstein warned Roosevelt that recent scientific discoveries made it possible to build a bomb more powerful than any weapon the world had known. If the United States did not begin work on an atomic bomb, Hitler's scientists might build one first.

Einstein's letter led to the Manhattan Project, a secret scientific effort that many other refugees contributed to. It produced the first atomic bombs, which the United States used against Japan in 1945.

Einstein continued making major contributions to physics when he was past age 70—a remarkable feat in a field in which most original work is done by young people. He died in 1955.

THE VOYAGE

Gottlieb Mittelberger described his journey to Pennsylvania in 1750, where he lived for four years before returning to Germany. He boarded ship in England for the voyage across the Atlantic.

During the voyage there is on board these ships terrible misery, stench, fumes, horror, vomiting, many kinds of seasickness, fever, dysentery, headache, heat, constipation, boils, scurvy, cancer, mouth rot, and the like, all of which come from the old and sharply-salted food and meat, also from very bad and foul water, so that many die miserably.

Johann Augustus Roebling, who designed the Brooklyn Bridge, kept a journal of his 11-week transatlantic trip to the United States in 1831, when he was 25. In it he described an encounter at sea with a pirate ship.

Sunday, 10th July. Yesterday we encountered a very suspicious ship. In the morning we noticed four vessels, among others a light-sailing schooner, which took its course...toward us. Supposing that the latter wished to speak to us, our captain made ready the [signaling] flags and megaphone.... The schooner passed by...without greeting us with her flags. On the deck we descried [saw] about twelve persons; two others were sitting up in the crow's nest, who seemed to be observing everything on our deck.... One of the crew, apparently the captain of the band, addressed questions to our ship through a megaphone in a coarse, unintelligible dialect, which were answered in the English language....

The schooner passed by, mocking us with its megaphone, whereupon...suddenly more persons appeared on her, which were not to be seen before. We proceeded undisturbed; but our captain believes that that ship was a pirate, but that it did not venture to attack our ship, because it discerned many passengers aboard and thus had to expect much resistance and little goods [because it was not a cargo ship].... As a rule the pirates do not venture so far up here in the Atlantic Ocean and cruise only in the West Indian and South American waters, where they have their lurking hole.

Armanda Fallier von Rosenberg sailed to the United States aboard the ship Franziska *in the 1840s. She traveled in style, as she wrote in a letter.*

We have bid Bremen farewell since yesterday and spent our first night on the ship.... The cabin is large, and alongside are the rooms, which can be locked, each of which sleeps four persons, two below and

These Germans are on their way to the United States in 1925.

two above. The bunks are sufficiently wide and long; only the height is lacking, but we will get used to that....

A little bell summoned us to the deck. The captain standing in front of a festively decorated pulpit, read a suitable song out of a Reformed Church hymnal and then read with a loud strong voice an uplifting poem entitled "The Storm."... The poem described the circumstances of the emigrants, from the weak to the strong man, the dangers of the ocean voyage, the storm and the rescue, the new and often sad conditions of the new home. It was so affecting that eyes were filled while all about us was the surging ocean, the swaying ship and the poorly clad folks from the middle deck who sat at the captain's feet with devout faces paled by seasickness.

Angela Heck came to the United States in 1854 with her husband Nikolaus. She wrote a letter home describing a terrible storm they survived on the trip.

Around 12 o'clock there was a noise up on deck that was so loud that everyone awoke. The sailors and helmsman and captain were all on deck for we were now having such a terrible storm that we thought the ship would be torn apart. We all started to tremble and shake.... At daybreak things got even worse, that is on Easter Monday. The ship was listing to one side and all the top planks started to break. We had to hold on as tight as we could to keep from falling out.... We all called on all the Saints in Heaven and God, the Holy Mother of God, Saint Nikolas [the patron saint of seamen], but things kept getting worse and worse. We thought the ship would be ripped apart at any moment. The ship was listing

Passengers board a ship of the Hamburg-America Line around 1910. Founded in 1847, the Hamburg-America shipping company was among the first to specialize in transporting immigrants, rather than cargo, across the ocean.

so much that you couldn't lie down, stand up or sit.... The stairs were then closed off since water was coming into the ship from the deck above and the small trunks started floating around. Our cooking pot and spoon floated all around.... It was so loud, the ship was sailing just like it was in a valley and on both sides it was so high you couldn't see over, just water everywhere. The ship started to crack, two masts broke and their sails and ropes were ripped and torn to pieces. Then the ship sank down very deep and water came into the opening like it was being poured in with a bucket. They couldn't shut this or else we would have suffocated. We were all so frightened we couldn't even pray. We repented our sins and we all prepared to die.

Wilhelm Bürkert departed from Hamburg on September 15, 1875. He described the ordeal of his voyage.

We got on a nice small steamship. In two hours we were out of the [River] Elbe. But here we were met by a big ship like you can't imagine, with two big smokestacks [to which the passengers transferred for the ocean voyage]. Everyone jumped for his mattress, assigned by the main agent at the inn in Hamburg, and for his tin ware. It took me an hour to find mine. When I found it, I carried it to the sleeping place I had picked out. But before I realized it, my chamber pot and water bottle had been stolen.

I reported to a steward that they'd been stolen; he said then go steal some yourself. I took one for myself and locked it up in my traveling bag, which was a great help. After 2 hours, the ship started out. We were given lunch.

I was all the way in the front in the first group of berths, I had to eat with 11 others, and carve the meat. Every day, rice soup, potatoes and meat, potatoes and rice soup, except for the 2 Sundays, when we had pudding along with it. Oh, I believe that stuff was boiled in sea water, it made me so sick, I ate nothing and drank nothing. In the morning all they had was black coffee, if you can call it coffee, then nothing until lunch mentioned above and then in the evening nothing but tea, which was just as bad as the coffee....

On Saturday the 18th of Sept., I collapsed on the deck. They took me to my berth, where I hit, bit, scratched and ripped all the clothes of the sailors who had to hold me down. They gave me chloroform to put me to sleep. The ship's doctor gave up all hope. On the following Monday they were all waiting for me to die.

Oh, how horrible it is when you don't have anyone. My heart was pounding. They put something on it, it was like fire. The doctor said if he hadn't done that, my heart would have burst.... It was Wednesday before I was allowed to leave the hospital [on board ship]. During this time I had food from the first cabin, where princes and dukes travel. So it must not have been seasickness since then you have to throw up.

An 1874 woodcut from Harper's Weekly *depicted German emigrants boarding a ship in Hamburg, bound for New York City.*

Rosina Riedlinger came from southern Russia to the United States in the early years of the 20th century. She described her trip.

We had taken fourth class berths on the ship which were very nice until the crowded rooms became full of seasick passengers. People vomited everywhere, and the ship's workers tried to keep it clean but it was not possible. Besides this smell, many of the passengers were so lazy they would not even wash themselves.... Some of the women on the ship were dirty from head to foot and had lice. When we reached America we had to wash our heads with sulphur to get rid of the lice.

Fred Martin came to the United States from a German village in Russia in 1909. He described the voyage to his niece.

Immigrant families were crowded everywhere, along with boxes and barrels of supplies. Everything smelled badly. Everywhere you turned, you bumped into someone or fell over a bundle. It was pure havoc. One or another of the boys was bawling most of the time, especially George. It took a lot of impatience for me to hit a child, but one night I had had it. In anger I got up, struck a match, and lit the kerosene lamp on the wall. My eyes focused on the ceiling quite accidentally and I saw a mass of crawlers squirming and creeping into crevices. I examined George's body and found bedbugs crawling about, his body covered with red blotches, and then I knew why he was crying.... We left the kerosene lamp burning all night, considered an extravagance in those days, but it kept the bedbugs away....

One day a storm hit the ship. For the boys, it was exciting, and before I realized it, they were on the top deck pushing each other into the massive waves that splashed onto the deck. I caught one of them just in time before he was washed into the sea. I was at one of my weakest points on the entire journey. I shivered from fright. I stood on the deck with my hands and arms outstretched shouting, "Gott, bring uns doch zum land" ("God, bring us yet to land").

Passengers aboard the SS Amerika headed for the United States in 1907. Between 1890 and 1920, a high percentage of German immigrants were industrial workers seeking better wages and jobs. Many of them left their families behind and intended to return to Germany.

Emigrants on a ship from Germany around 1900.

CHAPTER THREE

PORTS OF ENTRY

Because Pennsylvania welcomed German religious dissenters, Philadelphia was the most frequent port of entry for German immigrants during the colonial period. One German American citizen of Philadelphia described the arrival routine of an immigrant vessel in 1728: "Before the ship is allowed to cast anchor in the harbor, the immigrants are all examined as to whether any contagious disease be among them. The next step is to bring all the new arrivals in a procession before the city hall and then compel them to take the oath of allegiance to the king of Great Britain. After that they are brought back to the ship. Those that have paid their passage are released, the others are advertised in the newspapers for sale."

Philadelphia had no monopoly on German redemptioners. In 1709, the government of England encouraged several hundred of them to go to New York by giving them land north of the city in return for their labor. In the 1720s, the French government attempted to colonize the territory of Louisiana by inviting German settlers to New Orleans. For the rest of the 18th century, German immigrants stepped off the ships to begin their American lives in virtually all the colonial ports,

from Boston to Baltimore, Charleston, and Savannah.

After independence, two of the United States's major exports to Europe were cotton and tobacco. Much of the cotton was shipped from New Orleans to the port of Le Havre, France; tobacco frequently went from Baltimore to Bremerhaven, in northern Germany. To avoid returning home with empty vessels, ship captains took back emigrant passengers, most of whom were German. Sizable numbers of these new immigrants then moved up the Mississippi River from New Orleans or inland on the Baltimore & Ohio Railroad.

In 1843, the newly independent Republic of Texas invited a group of Hessians to establish a colony in Texas. The next year, about 150 families arrived in the port of Brownsville, on the Gulf of Mexico. After they founded the city of New Braunfels, in the central Texas Hill Country, Brownsville became the gateway for many other German settlers.

Thousands of Germans also took the long sea journey around the southern tip of South America to reach San Francisco during the Gold Rush of 1849 and over the next few years.

It was New York, however, that became the nation's principal port of entry for German immigrants, as for

all other European groups. Nearly a million Germans (and almost as many Irish) arrived in New York during the 1850s. In response, New York established an immigrant-receiving station at Castle Garden, a former theater on an island off the southern tip of Manhattan Island. There newcomers were screened for diseases and given information about jobs and lodging, to protect them against "runners" who lured unwary immigrants to boarding-houses where they would be fleeced of their savings.

Some of Germany's charitable organizations established offices in New York to help newcomers. As Germans left Bremen, for example, they would be given the address of the New York German Society in the city. There they could find German speakers who would advise them on the best routes to their final destinations.

In January 1892, the federal government opened a new immigration-landing station, at Ellis Island in New York Harbor. By that time the peak of German immigration, in the mid-19th century, had passed, but even so about 1.5 million Germans went through Ellis Island until its closing in 1954. By then the international airlines were carrying the majority of the new immigrants to the United States.

GERMANS IN COLONIAL AMERICA

Francis Daniel Pastorius led the first German settlement in Pennsylvania in 1683. Because their beliefs were so similar, the Germans who were Mennonites united with the Pennsylvania Society of Friends, called Quakers. Pastorius described Germantown in a leaflet written in 1700.

A "house blessing" written around 1785 by one of the German Mennonite settlers in Pennsylvania. The Pennsylvania Dutch— as they were known because English-speaking Americans changed Deutsche *(German) to Dutch— created many kinds of folk art.*

The air is pure and serene, the summer is longer and warmer than it is in Germany, and we are cultivating many kinds of fruits and vegetables, and our labors meet with rich reward.

Of cattle we have a great abundance, but for want of proper accommodation they roam at large for the present....

Although this far-distant land was a dense wilderness— and it is only quite recently that it has come under cultivation of the Christians—there is much cause of wonder and admiration how rapidly it has already, under the blessing of God, advanced, and still is advancing day by day. The first part of the time we were obliged to obtain our provisions from the Jerseys [New Jersey] for money, and at a high price; but now we not only have enough for ourselves but a considerable surplus to dispose of among our neighboring colonies. Of the most needful mechanics we have enough now; but day laborers are very scarce, and of them we stand in great need. Of mills, brick kilns, and tile ovens, we have the necessary number.

Our surplus of grain and cattle we trade to Barbados for rum, syrup, sugar, and salt. The furs, however, we export to England for other manufactured goods.

Conrad Weiser was a Palatine German who arrived in the colonies in 1710. As a young man he settled with other Palatines in the Hudson Valley of New York State. In later life he would serve the colonies in dealing with the Native Americans. He described the colonial living conditions.

Bread was very dear [expensive], but the people worked hard for a living, and the old settlers were very kind and did much good to the Germans, though some of a different disposition were not wanting. A chief of the Maqua [Mohawk] nation, named Quaynant, visited my father, and they agreed that I should go with Quaynant into his country to learn the Mohawk language. I accompanied him, and reached Mohawk country in the latter part of November, and lived with the Indians; here I suffered much from the excessive cold, for I was but badly clothed, and towards spring also from hunger, for the Indians had nothing to eat. I was frequently obliged to hide from drunken Indians. Towards the end of July,

I returned to my father, and had learned the greater part of the Mohawk language. There were always Mohawks among us hunting, so that there was always something for me to do in interpreting, but without pay.

In 1734, Johann Andreas Zwifler, an apothecary, left Salzburg, Austria, after the archbishop who ruled there demanded that all its inhabitants be Catholic. Upon settling in Georgia, Zwifler wrote to some friends in Augsburg.

My dear friends, I am writing these few lines to you together so that you may see that God is with me and has shown His power in me not only by leading me here fresh and sound and without any stumbling block, but also by showing me a place for my life's occupation where His honor is advanced and His name glorified. That everything goes well with me here, you can see from my yet facile hand. I am ready to admit that I am as well off here as I have been anywhere. Again and again I see here the blessing of God richly flowing over me, and perhaps much of it was due to the many thousands of "God thank you's" uttered by our dear Saltzburgers, who had many hardships on our journey, which lasted 8 weeks on the sea. But God blessed my limited medicine so that I cured them all and not one, either large or small, died, but all came to shore well. I learned more on this journey than I ever had before: I learned myself...[how] to bleed people [opening a vein and letting blood flow was a common form of medical treatment], for which God so led my hand that I was fortunate and made no mistakes. I have opened the veins of more than 20 people, including both of the pastors, who willingly placed themselves in my hands. We live here in a good land, for which God be praised. There is no scarcity of wild game, Indian chickens [turkeys], and partridges; and good fish are abundant. We can not give enough thanks to God, who has provided us with all kinds of victuals, and continues to do so. We have received beef and pork, peas, beans, rice, flour, salt, butter, cheese, pepper, all kinds of roots, beside 60 kinds of seeds, also more than 20 cows, 7 horses, also oxen, and await still more gifts. We can not marvel enough at all the many benefactions that have come to us...and we are given even more than the other inhabitants. In this country we find honey and turpentine. This is enough for this time.

Gottlieb Mittelberger, who lived in Pennsylvania for four years in the 1750s, described the plight of redemptioners.

When the ships have landed at Philadelphia after their long voyage, no one is permitted to leave them except those who pay for their passage or can give good security; the others, who cannot pay, must remain on board the ships till they are purchased and are released from the ships by their purchasers. The sick always fare the worst, for the healthy are naturally preferred and purchased

German immigrant John Peter Zenger began to publish the New-York Weekly Journal *in 1733. His criticism of the British governor landed him in jail, charged with libel. Zenger continued to write and passed his editorials to his wife through a hole in his cell door. A jury acquitted Zenger, producing the first victory for freedom of the press in the American colonies.*

David and Phila Franks, the children of Jacob Franks, a prosperous New York shipowner and merchant. In 1729, Jacob Franks became the first German Jew to be elected president of New York City's Jewish congregation.

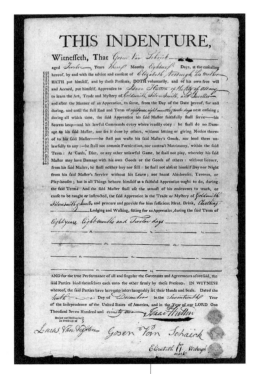

THIS INDENTURE,

Witnesseth, That Gosen Van Schaick
aged *Twelve* Years *Thirof* Months *Eighteen* Days, at the ensealing hereof, by and with the advice and consent of *Elizabeth Northingh his mother* HATH put himself, and by these Presents, DOTH voluntarily, and of his own free will and Accord, put himself, Apprentice to *Isaac Hutton of the City of Albany* to learn the Art, Trade and Mystery of *Goldsmith, Silversmith, and Jeweller* and after the Manner of an Apprentice, to serve, from the Day of the Date hereof, for and during, and until the full End and Term of *eighteen years eight months twelve days* next ensuing ; during all which time, the said Apprentice his said Master faithfully shall serve—his Secrets keep—and his lawful Commands every where readily obey : he shall do no Damage to his said Master, nor see it done by others, without letting or giving Notice thereof to his said Master—he shall not waste his said Master's Goods, nor lend them unlawfully to any :—he shall not commit Fornication, nor contract Matrimony, within the said Term : At Cards, Dice, or any other unlawful Game, he shall not play, whereby his said Master may have Damage with his own Goods or the Goods of others : without licence, from his said Master, he shall neither buy nor sell : he shall not absent himself Day nor Night from his said Master's Service without his Leave ; nor haunt Ale-houses, Taverns, or Play-houses ; but in all Things behave himself as a faithful Apprentice ought to do, during the said Term: And the said Master shall use the utmost of his endeavors to teach, or cause to be taught or instructed, the said Apprentice in the Trade or Mystery of *Goldsmith Silversmith &c.* and procure and provide for him sufficient Meat, Drink, *Clothing* Lodging and Washing, fitting for an Apprentice, during the said Term of

Eight years, Eight months and Twelve days

AND for the true Performance of all and singular the Covenants and Agreements aforesaid, the said Parties bind themselves each unto the other firmly by these Presents. IN WITNESS whereof, the said Parties have hereunto interchangeably set their Hands and Seals. Dated the *Sixth* Day of *December* in the *Seventeenth* Year of the Independence of the United States of America, and in the Year of our LORD One Thousand Seven Hundred and *ninety seven.*

Sealed and Delivered
In presence of

Lucas Van Scipolen   *Gosen Von Schaick*

Elizabeth H. Wittingh
marks

Many Germans in colonial times signed indentures, or contracts, to work off the cost of their ship passage. This indenture made Isaac Hutton, age 12, an apprentice who would learn the trades of goldsmith, silversmith, and jeweler.

A 1788 chest decorated in the style called Bauernmalerei (farmer's painting) may have been used by a young woman to store the quilts and blankets that she made before her marriage. Colonial-era Germans painted such brightly colored floral and geometric designs on furniture, clocks, woodwork, and plates.

first; and so the sick and wretched must often remain on board in front of the city for two or three weeks, and frequently die, whereas many a one, if he could pay his debt and were permitted to leave the ship immediately, might recover and remain alive.

The sale of human beings in the market on board the ship is carried on thus: Every day Englishmen, Dutchmen, and High German people come from the city of Philadelphia and other places, in part from a great distance, say twenty, thirty, or forty hours away, and go on board the newly arrived ship that has brought and offers for sale passengers from Europe, and select among the healthy persons such as they deem suitable for their business, and bargain with them how long they will serve for their passage money, which most of them are still in debt for. When they have come to an agreement, it happens that adult persons bind themselves in writing to serve three, four, five, or six years for the amount due by them, according to their age and strength. But very young people, from ten to fifteen years, must serve till they are twenty-one years old.

Many parents must sell and trade away their children like so many head of cattle, for if their children take the debt upon themselves, the parents can leave the ship free and unrestrained; but as the parents often do not know where and to what people their children are going, it often happens that such parents and children, after leaving the ship, do not see each other again for many years, perhaps no more in all their lives.

It often happens that whole families, husband, wife, and children, are separated by being sold to different purchasers, especially when they have not paid any part of their passage money.

When a husband or wife has died at sea, when the ship has made more than half of her trip, the survivor must pay or serve not only for himself or herself, but also for the deceased. When both parents have died over halfway at sea, their children, especially when they are young and have nothing to pawn or to pay, must stand for their own and their parents' passage, and serve till they are twenty-one years old. When one has served his or her term, he or she is entitled to a new suit of clothes at parting; and if it has been so stipulated, a man gets in addition a horse, a woman, a cow.

John Hayn, at one time a cowherd in the Nassau region of Germany, wrote home trying to convey the largeness of America.

People sell out and move on. Nobody knows yet how large the country is. On a map Europe would be like Dillenburg [a small town] compared to Frankfurt [a large city]. The best land is still lying waste. They travel some 700 miles and yet do not find the end.

Johannes Schlessmann, living near Germantown, Pennsylvania, wrote home in 1753.

I and my children and my wife thank and praise God a thousand times, that we are in this healthful country. We expect to support ourselves much better here than in Germany, for this is a free country.

Many Germans, at home and in America, practiced the trade of printing. Christopher Sower, or Saur, was the greatest German American printer of the 18th century. He printed the first Bible in a European language in the New World. His son Samuel was born in Germantown, Pennsylvania, in 1767 and entered the printing trade in Philadelphia. He then moved to Baltimore, where he had a print shop with a partner, William Gwinn. Samuel Sower wrote of the trials of his business to his brother in Philadelphia.

I am chained down to business closer than ever.... My partner will not bother himself with business, having invested between $7000 and $8000 in the business and built for me a home costing at least $3000. I see him not more than once a month and he leaves everything in my hands to manage, saying if he had not the utmost confidence in me, he would not have gone into it. The business of type founding [making metal type] is making great strides. Orders are pouring in from everywhere so we can not fill half of them. We have undertaken to cast the smallest types that have been used in the world.... We have eleven boys and six journeymen at work and orders for 5000 pounds of type. I am working night and day.

German Americans served on both sides in the American Revolution. As soldiers and gunsmiths, the German settlers played an important role. In addition, more than 30,000 mercenary soldiers (many from Hesse), whose services were sold by their rulers, served the British. After the war many of these Hessians decided to stay. A Hessian officer named Andreas Wiederholdt was captured at the Battle of Trenton and became a prisoner of war. He settled in Fredericksburg, Virginia, and recorded some observations on the local social situation in his diary.

The residents are the friendliest and most courteous in America no matter what their standings and their opinion. Tories and [Whigs] are hospitable and obliging to everyone, especially strangers. The women are pretty, courteous, friendly and modest...; we enjoyed a great deal of civility from them and not-withstanding the fact that we are enemies, they gave us a great preference over their own men at balls and other occasions.

Molly Pitcher

Mary Ludwig was born on a Pennsylvania farm around 1755, the daughter of German immigrants. She would become a legendary figure in U.S. history, better known as Molly Pitcher. When the American Revolution began, Mary was married to a man named John Hays. When he joined George Washington's army, Mary accompanied him. Like many other Patriot women, she cooked, did laundry, and cared for the wounded in camp.

On June 28, 1778, Washington's troops faced an army under British general Henry Clinton near Monmouth, New Jersey—the first battle for the Patriot army since autumn. All that winter, camped at Valley Forge, Washington had struggled to keep his army from starving. Clinton expected an easy victory. But during that time, Baron Friedrich von Steuben, a former Prussian officer, had been training Washington's army, turning it into a disciplined military force. The Battle of Monmouth would be its first test.

John Hays was assigned to an artillery gun. It was a brutally hot day, and the gunners were soon covered with sweat and weak from thirst. When his wife saw the situation, she filled a pitcher from a stream and carried it across the battlefield to bring water to the gunners. The troops dubbed her Molly Pitcher.

Mary's husband collapsed from the heat. Immediately, she took his place at the gun. As night fell, Clinton ordered his army to retreat. Washington's forces proved they could hold their own against a professional army.

The story of Molly Pitcher's bravery spread, raising the spirits of the rebels. But Mary's husband died the following year, and she returned home to Carlisle, Pennsylvania. In 1822 the state of Pennsylvania awarded her an annual income of $40 for her service to her country.

LANDING

In August 1831, Johann Augustus Roebling's ship landed at the port of Newcastle, Delaware, stopping for a health inspection of the passengers. Later it proceeded up the Delaware River to Philadelphia. Roebling described the inspection at Newcastle.

Saturday, 6th August. This morning we received a visit from the Health Officer, whose business it is to examine the passengers and crew of each newly-arrived ship...and thereby to prevent the introduction of contagious diseases. The physician, a fine gentleman...closely observed all the passengers and counted them several times, and asked if any persons had been sick at sea or had died. Only after the physician has been on board and all have been found well, are the passengers permitted to go ashore.

Theodor Engelmann fled Germany after taking part in a student rebellion in 1833. He described his arrival in America.

It was night as we entered the harbor of New York, a bright starlit heaven was above us. From the city and from the houses along the shore there glittered gas and lamp light...and distant barking of dogs interrupted the solemn stillness of the night—to our no little astonishment—since there ruled at the time the erroneous idea that the American dogs could not bark, but were mute. The next morning the sun rose in a cloudless sky and illuminated with her magic light the bay with its charming shores and the city in the distance.... "The Christkindchen [Christ child] has given presents," shouted a

Baltimore was the second most common port of arrival for German immigrants in the 19th century. The group here is being processed at Locust Point, the Baltimore equivalent of Ellis Island. From there, many took trains headed west.

German woman from within the ship as she was coming on deck. We had to spend one day in painful quarantine in sight of the long expected landing before we were allowed to set foot on firm land and to go into the city of New York.

Henriette Geisberg Bruns came to the United States from Westphalia in 1836 with her husband and young child. In a letter to her brother Heinrich she described her arrival.

Great was everybody's joy when the pilot ship approached us at Cape Charles [Virginia].... We were sincerely happy to have firm soil under us and hurried here [Richmond, Virginia]. We immediately went and brought fruit and everything tasted magnificent. We spent the night here. On Sunday morning we went with Hermann to the Cathedral. I was really moved when I found myself in the really beautiful church, and I was able to tell the Good Lord so very much from the bottom of my heart in a few minutes.

In 1854, Franz Joseph Ennemoser came to the United States from Mainz with his family.

Friday, December 8. Land! Land! We have finally sighted land! The passengers throng the deck, wild with joy. Many are throwing their sacks of straw into the water, heedless that they will have to spend another night on board.

By afternoon we are riding at anchor in the harbor of New Orleans. We have been 43 days at sea. We are all amazed at how green the country looks. Back home in Germany at this time of year, the ground is covered with snow. Here in New Orleans summer never ends. Everything is green, green, green! Each of us talks about the first thing he will do when he sets foot on shore tomorrow. My eight-year-old daughter, Maria, says that she will be happy to have only a glass of clean fresh water!

In 1875, 16-year-old Wilhelm Bürkert emigrated to the United States from Waldenburg, Württemberg. He wrote to his family about his arrival in New York.

At 4 in the morning we heard land—land. And that is a sight, oh splendid. We were in [Staten Island]. Here the anchor was cast. A doctor came out about 7 o'clock and examined each one to see if he had a contagious disease. Everyone was healthy. We were allowed to go on. We went to the town [of] Hoboken. Here we had to go on a small ship. And everyone's trunks were inspected to see if anyone was trying to smuggle something into America. There was a man who had to pay 40 dollars for 2 silk dresses and 1 gold mark. At 3 o'clock in the afternoon we arrived in New-York.

Here we all went into a garden where a speech was held. It was Castle-Garden, which is set up to take care of the emigrants. Innkeepers came to offer their services. We went to the inn "Zur Stadt Balingen" [Town of Balingen], the innkeeper is a Württemberger from Balingen.

German immigrants at Battery Park, on the southern tip of Manhattan Island, in 1896. A ferry from Ellis Island dropped them off there.

Jacob Mithelstadt and his family were Germans from Russia who arrived at Ellis Island in 1905. They settled in Kulm, North Dakota, where many Black Sea Germans lived.

Three Klein brothers...came to Texas from Germany. The first one kept his name unchanged and spelled it as it is spelled in German: Klein. The second one anglicized the spelling to Cline, and the third translated his name to Small [which is what Klein means in German]. Years later when a large inheritance came from Germany, Mr. Klein got all of it, and Mr. Cline and Mr. Small were left out completely. Father's moral was: do not renounce your heritage.

A 19th-century immigrant boardinghouse in New York City provided new arrivals a place to stay until they could find jobs and living quarters elsewhere. Unwary travelers, however, could be fleeced of their savings by cardsharps and other swindlers.

FINDING A PLACE TO LIVE

Caroline Ernst von Hinueber came to Texas with her parents in 1831. The family first landed in New York, where her father, Frederich Ernst, met a countryman named Fortran who accompanied them. Caroline Ernst's family had originally planned to settle in Missouri. But after arriving in New Orleans, they heard about the Mexican government's offer of land to settlers in Texas. She wrote about her journey there.

We set sail for Texas in the ship *Saltillo*. Just as we were ready to start, a flatboat with a party of Kentuckians and their dogs was hitched onto our vessel. The Kentuckians came aboard, leaving their dogs behind on the flatboat.

We were almost as uncomfortable as the dogs. The boat was jammed with passengers and their luggage so that you could hardly find a place on the floor to lie down at night. I firmly believe that a strong wind would have drowned us all. We landed at Harrisburg [near Galveston Bay], which consisted at that time of about five or six log houses, on the 3rd of April, 1831.

Here we remained five weeks, while Fortran went ahead of us and obtained land, where now stands the town of Industry....

After we had lived on Fortran's place for six months, we moved into our own house. It was roofed with straw and had six sides, which were made out of moss. The roof was by no means waterproof, and we often held an umbrella over our bed when it rained at night, while the cows came and ate the moss. Of course we suffered a great deal in the winter. My father had tried to build a chimney and fireplace out of logs and clay, but we were afraid to light a fire because of our straw roof. So we had to shiver. Our shoes gave out, and we had to go barefoot in winter, for we did not know how to make moccasins. Our supply of clothes was also insufficient, and we had no spinning wheel, nor did we know how to spin and weave like the Americans.

After landing in Baltimore in 1836, Henriette Geisberg Bruns, her husband, young son, and two brothers set out for Missouri. In her autobiography she recalled the trip.

The *Cavalier* [a riverboat they had boarded at Cincinnati] brought us to St. Louis without trouble. We arrived there in rainy weather. The city was overcrowded with people traveling home to the south.... There we purchased the most necessary things for our immediate needs, and then we traveled on the Missouri up to Jefferson City.... We went up the

Osage River [and] disembarked at Henry Dixon's inn.... There our people found some shelter until the new friends from Westphalia [Missouri] came to get us. I was shocked at their appearance, genuine backwoodsmen in rough clothing and prepared for rough work.... It was certainly a great omen that they welcomed us in this manner. And they probably had very little choice. A wagon covered with a tarpaulin had arrived. Mr. John Shipley took us through thick and thin, often without any road, to his parents' farm on the Osage River. The next morning we were ferried across the river, and after a few hours we were at our new home. It was the second of November 1836....

Our first shelter was in the middle of a field, a simple log cabin with two bedsteads, one table, four chairs, and one bench. We were longing for rest and were not even surprised at the sparse furnishings.

After the Bohning family disembarked in Baltimore in 1843, their journey was only half over. They were headed for Cleveland, where a relative lived. Other immigrants on their ship also had destinations inland. Ernst Bohning, who was 10 years old at the time, later described the long overland journey, which began with a short train ride. After that, they spent many days on a canal boat.

I first looked at the railroad cars that would haul us. They were small, light, flat cars fitted with little wheels. We sat on rough-sawn benches, arranged crossways, half of the passengers riding forward, and the other half, backward. To protect us from the sun and rain, a board roof had been crudely knocked together.

I also wanted to see the little iron horse that was supposed to pull us. It was spare and small. The belly was like a sugar barrel; it stood on end rather than on all fours; the neck was like a stovepipe, standing straight up. There was no head, and I didn't notice any tail. This was how the little horse looked, and I was worried that it could not pull us with all our chests

A group of German immigrants arrives at the railroad station in St. Paul, Minnesota, in 1902. They wore the typical clothing of Alsace, their native region.

As Others Saw Them

Frederika Bremer, a Swedish writer, published a diary of her travels through the United States in the 1850s. She wrote of Milwaukee:

Nearly half of the inhabitants are German, and they occupy a portion of the city to themselves, which is called "German Town." This lies on the other side of the river Milwaukee. Here one sees German houses, German inscriptions over the doors or signs.... Here are published German newspapers, and many Germans live here who never learn English and seldom go beyond German Town. The Germans in the Western states seem...to band together in a clan-like manner, to live together and amuse themselves as in their fatherland. Their music and dances, and other popular pleasures, distinguish them from the Anglo-American people, who, particularly in the West, have no other pleasure than business.

Women and children from the Volga River region of Russia step off the train in Lincoln, Nebraska.

and trunks. Although it was small and slight, it had an unruly spirit for inside it rumbled and roared a lot. I became so worried and frightened, I ran away after seeing the driver who would drive the little iron horse....

Then the journey got started, at first slowly, then faster and even faster. The little iron horse pulled well, and we were pleased, for the gentle rolling along on tracks really seemed like Paradise after the hardships of the ocean journey....

We had been well provided with food, but had no beverages with us, and the crew of the train was not concerned about that; we didn't stop till night. We were in a clearing in the woods on a wonderful, moonlit summer night. There was no station in sight, but the train stopped and the iron horse drove on, leaving the train with us emigrants there in the woods. We didn't know what to make of this, but it didn't bother us much, and we all climbed out to stretch our stiff legs and move around.

Then we discovered that the standing timber at the edge of the clearing had been cut down, and the brush was piled up in heaps. It was good and dry, and we lit a heap which soon blazed brightly. At the fire, we prepared a decent dinner and fixed German pancakes. Those who had pans were in good shape, and the others who had no pans, borrowed pans later, and by midnight, everybody had eaten....

That was the first day, and it was followed by twenty-five more days of travel in America.... We continued in a canal boat pulled by two miserable horses.... You can imagine how we were squeezed together, and how we had to travel in a terribly narrow space. By day it was alright; the men walked on the shore beside the boat and hunted for food when they had the chance. We young folk also went on foot and looked for birds' nests or something to eat. But at night! The boat was much too narrow to give us even a minimum of space to lie down on the deck....

[Finally they arrived in Cleveland.] But it wasn't until the next morning that we could greet our brother and Ernst Borges, who lived six miles out from the town.... I don't need to describe how happy we were to be all together again, safe and sound. Now we forgot all the troubles and suffering we had endured on the long journey. That lay behind us and we looked cheerfully and with renewed energy to the future.

Cornelius Duerksen, a Mennonite fleeing the military draft in Russia, came to the United States in 1874. He described his trip to Kansas in his daybook.

September 4, Friday, at 5:30 P.M. we were leaving Kestel garden [Castle Garden] by ship to the R.R. depot for our trip to Kansas.... Saturday, September 5, last night the beginning of our trip to Kansas, we saw that the land was mountainous and covered by forest. By evening we came to Bauvallo [Buffalo], where we crossed the...Angere [Niagara] River. The bridge is 118 feet high and 828 feet long.... We wondered...would it ever end? We heard the water falls....

September 6, Sunday, we traveled safely during the night and morning, till 12:30 noon, when we reached Detroit. We were taken by a ship over an arm of Lake Huron. At 2 P.M. we left for Chicago.

September 7, Monday, at 6 A.M. we arrived at Chicago, and left at 8 A.M.... Chicago is a large inhabited city, almost as large as New York. We had to change cars. They are much nicer and comfortable. The seats are for two, they are also upholstered. There is a large window at each seat. The windows can be opened to let in fresh air. The cars seat for 54. There are two iron stoves. There is a water container which is filled with water and ice, and replenished when it gets empty. The drinking cup is a tin cup fastened with a small chain. In one corner also is a restroom. This is the only train equipped like this from New York to Kansas for immigrants. After leaving Chicago, we saw very plain land, and now it is hilly again, and we see many different kinds of trees, also we saw many fruit orchards. They were so beautiful. They [are] covered with delicious fruit. Nice horses, beautiful horned cattle, and hogs were on freight trains coming from the west, going east to the slaughter houses. Also horses were shipped by train.

September 8, Tuesday, at 9:30 A.M. we arrived at Atchison, the border of Kansas. Here we left the train. There were several wagons, some drawn by 2 and some by 4 horses, to take us to the other train. We had to cross the wide river on a large steamboat, then we came to the R.R. station, where the train was waiting for us, and we boarded the train at once.... Left at 12:30 and arrived at Topeka at 4 P.M. We were taken to a large empty manufacture building, on the end of it, we all were to lodge here. We were given boards to put on the ground (it was very dusty) to sleep on and to put our belongings. In the evening a number of us went to the R.R. station to inquire about the price of the land.

Sophia Kallenberger Beck was born in 1877 in Neuburg in southern Russia and came to America with her family at age nine. She told an interviewer in 1939 about their trip inland from New York.

Mother's health was so frail and the severe strain of this voyage exhausted her so completely we were compelled to remain in New York about two days while she recuperated sufficiently to resume the journey. During this time we remained in a large single room structure [probably Castle Garden]. After a two-day rest, we resumed our journey to Scotland, Dakota Territory, but at Marion Junction the whole train was delayed three days due to the illness of Mother and others, because of which the authorities feared an epidemic. During this time father had to beg food from the town people for his family because these unexpected delays had completely absorbed all his funds. About six days after leaving New York we finally arrived in Scotland, at the house of my sister Catherina, Mrs. Wagner. Father had only 25 cents in his pocket when we arrived.

John Jacob Astor

Arriving in Baltimore in 1784 with seven flutes and $25, John Jacob Astor would in time become the richest man in the United States. Born in the Rhineland village of Waldorf in 1764, Astor left home at the age of 17. He worked in London for three years at his brother's musical instrument shop. Another brother was a butcher in New York, and Astor decided to seek his fortune there.

On the ocean voyage, Astor learned about the fur trade from a fellow passenger. Within a year he was traveling back to London with furs for sale. He used his profits to invest in farmland in Manhattan, foreseeing that the city would expand northward. In 1811, Astor financed a voyage to the far Northwest to find new fur sources. The expedition members founded a trading post on the site of today's Astoria, Oregon. Soon Astor's ships were crossing the Pacific to trade with China.

Astor enjoyed music and literature. He paid Washington Irving, the leading American author of the day, to write *Astoria,* a glorified history of his patron's western trading empire. In his will, Astor set up a public library to house his extensive book collection. Later it was combined with two others to form the New York Public Library.

By the 1820s Astor was known as "the landlord of New York" because he owned so many buildings. He told his son never to sell land but to lease it so that the profits would keep coming. The advice was followed until the 20th century, when Vincent Astor, John Jacob's great-great-grandson, sold the land on which the Empire State Building now stands.

In his later years, Astor's health began to fail, but he insisted on keeping track of business. To give him exercise, servants would toss him gently into the air on a blanket. During such a session, one of Astor's rent collectors arrived and reported that an elderly woman could not pay her rent. From the bouncing blanket, Astor said, "I tell you she can pay it, and she *will* pay it." On the way out, the rent collector met Astor's son, who secretly paid the rent. The old man was gleeful. When he died in 1848, Astor left an estate of $20 million, the largest fortune in the United States at that time.

FIRST IMPRESSIONS

These two pictures appeared in a book published in Germany in 1838. Titled The Dream *(top)* and The Reality, *they were intended to warn would-be emigrants that life in the United States was not as easy as they imagined.*

In his journal, Johann Augustus Roebling gave his impressions of Philadelphia in 1831.

Nowhere does one see a person in rags; all, even the common workmen, go very cleanly and neatly dressed. A German workman [in Germany], who works in only shirt-sleeves, appears like a chimney-sweep alongside one of these.... I have spoken to no one yet, even when he seemed to be in a hurry, who has not fully replied to my question. Every American, even when he is poor and must serve others, feels his innate rights as a man. What a contrast to the oppressed German population!

Hermann Seele, born in the kingdom of Hanover in 1823, arrived in Galveston, Texas, in December 1843. He recalled his first Christmas Eve in the United States.

I had a great desire to see, this very evening, a Christmas tree. Finally, my wish was to be fulfilled. On the right side of the street stood a house in a yard with fragrant, blossoming rosebushes.... I stood a while in front of the garden gate, watched and listened to the muffled sound of happy children's voices.

Finally, I ventured hesitantly to open the gate and approached cautiously with soft steps over the sandy path.... At the nearest window, I peered through the slats of the shutter...and looked into a cheery room. There, in the splendor of a small pyramid of lights decorated with green branches and brightly colored ornaments, I saw the eyes flashing with joy and the cheeks glowing with pleasure of the dear little children with their magnificent pile of gifts.... Happy and yet full of melancholy too, I went quietly back to the gate as unnoticed as I had come. From there I walked on through the dark night, a lonely stranger in a country still so new to me.

Dorothea Klein came to St. Louis from Prussia in the 1850s to join her husband. She wrote home of her first impressions.

This land and this town with the African heat...makes many demands. German underwear and clothes...are totally unsuitable for the climate at this time of year.... In spite of opening the shutters and creating drafts, the air remains sultry and the moskitbars [an anti-mosquito device] and bed hangings...seem to choke one. The mosquitos are much bigger than gnats in Germany, demand their rights in the evening and at night and we all have great respect for their singing and their cunning.

Many German immigrants were impressed by the abundance of the United States. An immigrant wrote home in the 1870s about the quality of his boardinghouse food in New York.

Beefsteak. Amazing! Meat in the morning! It is really nice in America! On the other side I only had coffee. And then lunch that was even more delicious! Soup, two kinds of meat, different kinds of vegetables—I only had one of a kind on the other side—steamed fruit and last not least a glass of beer after all this! This was fun for me, I must stay here, where could it be better?

Sister Reinhardt Hecker came to the United States in 1914 as a teenager. At Ellis Island she ate a surprising new food.

Before we left for the train they gave each one a big box of bananas and apples and sandwiches. Each child got one. I don't know why they did that, but anyhow we didn't know how to eat bananas. So we'd bite into it, [and say] oh they're no good, they're no good. So on the way up I saw some people, a couple sitting there and they peel those bananas. I says, "You're supposed to peel them." Then we learned how to eat bananas cause we never saw those in Europe.

The movie and play The Sound of Music *were based on the autobiography of Maria von Trapp. She and her family escaped from Nazi-held Austria in 1938 and came to the United States. Von Trapp once described her first encounter with an escalator, in Macy's department store in New York City.*

When I saw people step on it and be moved upwards in front of my very eyes, I got an uncomfortable feeling, as though I were witnessing witchcraft. When invited, however, to take the fatal step myself, I most vigorously declined. Meanwhile, a number of people had gathered behind us. We were blocking traffic.

"Go ahead, don't be silly," whispered Georg [her husband] encouragingly. With a lump in my throat, I put out one foot hesitatingly, but when it touched that moving thing, I quickly drew it back as if I had been bitten by a snake. Now good-natured, kind-hearted Americans gathered around me, and good advice from all sides made the situation more embarrassing by the second....

"Close your eyes, lady, and take a step."

This was the best advice given so far. Now I was on it. How would I get off? That is easier. The staircase just slides you off, whether you want to or not.... Up to this day whenever I have to take an escalator, I close my eyes and take a deep breath.

A German immigrant in traditional clothing at Ellis Island around 1885. His lederhosen (leather shorts), embroidered suspenders, and green hat were actually worn only in Bavaria. But today they are donned for ethnic celebrations by many German American men.

A German American woodsman and his child in 1917. German Americans typically kept the land they settled within the family for generations. Even today, they are the largest American ethnic group working in agriculture.

CHAPTER FOUR

A NEW LIFE

Most of the German settlers who arrived in Pennsylvania in 1683 and established Germantown were cloth weavers. However, numerous other craftsworkers and artisans arrived from Germany in colonial times. Printers, bookbinders, papermakers, carpenters, cabinetmakers, blacksmiths, tailors, cobblers, ironworkers, and stonemasons found a market for their skills in the English colonies. Silkworkers from the town of Neufchatel established a colony in Beaufort County, South Carolina. There they raised silkworms on mulberry trees planted on 40,000 acres.

German redemptioners sometimes learned trades as apprentices. Such was the case with John Peter Zenger, who arrived in 1709 at the age of 13 and spent eight years under contract to a printer in New York. Zenger later founded his own newspaper, *The New-York Weekly Journal*. A libel suit brought against him by the colonial government resulted in the first legal victory for freedom of the press in the American colonies.

German merchants also set up shop in the New World. Some established taverns stocked with beer, the favorite German beverage. Immigrant Germans founded

breweries in New York and Baltimore in the early 1700s.

The majority of the colonial German immigrants were farmers. Though the American land seemed limitless, much of it—at least in the English colonies—was covered with forest. Starting a farm meant chopping down trees, clearing the land, and digging out rocks that stood in the way of plows. Undaunted, the German immigrant farmers moved farther into Pennsylvania, up the Hudson River in New York, and into northern New Jersey. Fewer Germans went to New England, though some settled in the Broad Bay region and along the Kennebec River in what later became the state of Maine. German colonial farmers also settled in Virginia, Maryland, Delaware, the Carolinas, and Georgia. Those who were brought by the French into Louisiana as settlers moved up the Mississippi and also into the present-day state of Mississippi.

Eighteenth-century German farmers in the Conestoga Valley of Pennsylvania developed a deep-bodied covered wagon to take their crops to market. In the next century, Conestoga wagons modeled after these took thousands of pioneers across the western plains.

In the mid–1800s, German American farmers continued west-

ward across the continent. Many took advantage of the free public land offered by the Homestead Act of 1862. From Ohio to Indiana, Illinois, Missouri, and Iowa they planted corn, a crop seldom grown in Germany. A new German immigrant to Missouri in 1861 wrote home: "Corn...that's the most important thing in America, man and beast live from it." Indeed, much of the corn was of a type specifically designed for feeding animals, chiefly the pigs that were among the products of farms in the "corn belt."

The largest number of German Americans took up dairy farming. The "dairy belt" included parts of upstate New York as well as Wisconsin, Michigan, and Minnesota. Dairy farms also tended to cluster around large cities, so that their cheese, butter, and milk could be rapidly delivered to urban markets.

The cultivation of wheat was a specialty of Germans from Russia. In 1872 the Russian government revoked the special privileges originating with Catherine the Great that had drawn German immigrants to the Volga River and Black Sea regions in the previous century. The action cost Russia some of its best farmers, as thousands of German settlers migrated to the United States. They brought with them the seeds of hard Turkey red winter wheat. This type of wheat could be planted in the fall and survive the harsh winters of the northern plains states for spring harvest. Germans from Russia sowed this crop in the Dakotas, Nebraska, Kansas, and parts of Colorado, helping to turn the vast grasslands into wheat fields that became the breadbasket of America.

Germans from Russia also put down roots in California, planting grapevines. Those from the Volga region settled around Lodi in central California, and Black Sea Germans formed communities in the San Joaquin Valley around Fresno, where they helped establish a raisin industry.

Many German immigrants in the first half of the 19th century were university graduates. Some found jobs as teachers, journalists, and clerks, but others tried their hands at farming. They earned the nickname Latin farmers, after their classical training in ancient Latin and Greek, which ill suited them for the hard life of farming.

Despite the enormous influence German American farmers had on U.S. agriculture, a majority of German-speaking immigrants engaged in other kinds of work. Some became legendary success stories. John Jacob Astor, who arrived from Waldorf, Germany, in 1784 as a teenager, became the United States's first millionaire from involvement in the fur trade and real estate investments. Frederick Weyerhaeuser, who arrived penniless in 1852, started work in a sawmill; by 1900 his

lumber company owned almost 2 million acres of land. Heinrich Steinweg took his family to New York in 1850 and opened a piano business that became known as Steinway; its products are still standards of musical excellence.

Brewers of beer became wealthy and prominent members of every large German American community. The Pabst and Schlitz families in Milwaukee and the Busch family in St. Louis used their fortunes to build parks and other public facilities in their communities.

More than most other immigrant groups, German Americans found jobs as skilled workers. Many had learned trades in their

On the plains, trees were scarce, so industrious farmers built houses, like this one in North Dakota, that were made of sod cut from the prairie.

native land. Bakers, butchers, brewers, tailors, barbers, carpenters, cabinetmakers, and gardeners did not have to start at the bottom as low-paid unskilled laborers. For instance, the U.S. printing industry was dominated by German Americans, carrying on the tradition begun by Johannes Gutenberg in 1452. Artists, musicians, and clergymen were also among the German immigrant population.

The relatively high-paying jobs

of German American men enabled their wives to remain at home to fulfill their traditional roles as mothers and homemakers. Relatively few German American women entered the labor force, except as teachers and domestic servants. In 1890 about one out of every five German-born women in the United States worked as a maid, housekeeper, or cook. Many German American women found teaching a rewarding profession. Margaretha Meyer Schurz, wife of the '48er Carl Schurz, is often credited with starting the first kindergarten in the United States.

The great numbers of German and Irish immigrants who arrived in the United States in the 1840s and 1850s resulted in a wave of antiforeign prejudice. "Nativist" speakers argued that these newcomers were taking jobs from native-born Americans. The fact that so many of the newcomers—nearly all the Irish and about half of the Germans—were Roman Catholic caused fears among those who regarded Catholicism as a threat to American traditions.

In addition, German immigrants generally continued to speak their native language, which set them apart from the majority of Americans, who spoke English. And in many towns, Germans' consumption of beer on Sunday, often in lively beer halls, brought condemnation from Anglo-Saxon Protestants who felt this to be a violation of the Sabbath.

In 1845, a group later known as the American Party was founded to block foreign immigration. In the election of 1854, this party reached the height of its influence by electing governors or a majority of the legislatures in seven states. Soon afterward, however, the American Party split over the issue of slavery.

In 1856 the newborn Republican Party made an appeal for German Americans' votes by publishing its antislavery platform in German as well as English. Four years later, the Republicans' Presidential candidate, Abraham Lincoln, won a close election with the strong support of German American voters in key states.

During the Civil War, many German Americans served enthusiastically in the Union army. Fewer fought in the Confederate army, for the majority of German Americans lived outside the Southern states. Germans in St. Louis formed a militia that helped ensure that border state's loyalty to the Union. Numerous German Americans rose to the rank of general, including Carl Schurz and the flamboyant George Armstrong Custer, whose great-grandfather (named Küster) had been one of the Hessians who stayed in the United States after the Revolution.

The latter half of the 19th century saw the rise of labor unions and social reform movements in the United States. Labor unions had been formed in Germany as early as the 1840s, and German immigrants played an active role in the U.S. union movement. Skilled German American workers like bakers, tailors, and cigarmakers formed local trade unions in cities such as New York, Philadelphia, Milwaukee, St. Louis, and Chicago.

Many German Americans, inspired by the ideas of the German philosopher Karl Marx and other European socialists, saw the labor movement as part of a larger social transformation. The socialist movement was particularly strong in New York and midwestern cities with large German American

German brewers brought their skills to the United States and established countless breweries. This group posed outside a brewery in Milwaukee, which in 1900 had a larger proportion of German Americans than any other major U.S. city.

populations. In 1916 the mayor of Milwaukee and 21 of the 25 members of its city council were socialists.

By today's standards, the goals of the 19th-century socialist labor leaders were modest ones. German Americans led the fight for an eight-hour workday, retirement benefits, and disability insurance. However, business leaders condemned such demands as radical ideas, and bloody clashes between strikers and police turned public opinion against the socialist movement. The more conservative American Federation of Labor (AFL), established in 1886, eventually drew most of the German American trade unions into its membership. However, the AFL refused to admit unskilled workers in such industries as mining, construction, and manufacturing.

In the 20th century, the growth of assembly-line industries such as automobile manufacturing created a new wave of labor organizing.

Walter Reuther, a German American born in West Virginia in 1907, became active in the struggle to organize automobile workers during the 1930s. Reuther served as president of the United Automobile Workers (UAW) from 1946 until his death in 1970. In 1955, he led the merger of the Congress of Industrial Organizations (CIO), an association of industrial workers, with the AFL.

Socialism, which was so strong an ideal among 19th-century German Americans, failed to attract the support of most other Americans. Reuther himself fought off a communist attempt to take control of the UAW. Yet many of the German socialists' goals have won universal acceptance: the high standard of living enjoyed by most American workers, the Social Security program that enables elderly Americans to retire comfortably, and government-enforced safety in the workplace.

FARMING AND HOMESTEADING

Many German immigrants dreamed of starting a farm on land of their own. Franz Joseph Löwen came to the United States in 1857 and settled in Michigan. In a letter sent home in 1883, he described the importance of machinery to farming in the new land.

Brother Heinrich asked in his last letter how the land is worked. Nothing here is dug by hand like over there, everything is plowed and harrowed with horses or oxen. Potatoes, corn and such are done with a 2-3 or 5-shovel so-called cultivator pulled by a horse to get out the weeds, they are hilled up by a shovel plow. For every kind of work in the field there are machines but it often takes a long time before you can buy them because they are very expensive and a new farm that is only partly in cultivation often doesn't yield more than what one needs for one's own use. Then little by little, as his income increases, a farmer buys the machines he needs. In general all the farm machines here are very light and elegant but very strong. And you will find it hard to believe that I cut 22 acres of grain and meadow alone, and by hand.

But the strength of my younger years has passed somewhat and so for the last season I bought myself a reaper for the price of 120 dollars. I can reap 5 to 8 acres of grain a day, then I also have a 10 foot wide hay rake so that I can rake up all the hay and leftover grain with a horse. Now I only need a mowing machine for grass which costs 80 dollars, then later perhaps a sowing machine. But for now I am happy not to have to do the hardest work by hand any more.

Johann Witten, who arrived in the United States with his wife in 1882, began farming on rented land in Illinois. Five years later, now with two young children and a third on the way, Witten wrote his brother in Hanover, describing his decision to move west to buy his own farm.

The Scheck family of Kansas. In 1900, about 85 percent of all German Americans lived in the midwestern states from Ohio to the Rocky Mountains.

Joseph and Annie Burkholder and their family in 1901 on the day they moved from their sod house to a large two-story home (not seen here) in North Dakota.

We are going to leave here in the spring, I don't know yet where our home will be. I am going to go far away, where land is still cheap. [He would go to the state of Washington.] Up until now I have only rented it. I don't want to do that any more, I want to have my own land. I have to sell most of the things that I have. I have 26 head of cattle, they'll bring quite a bit of money. We had 9 milk cows this summer. From May 1st until now we have made almost 90 dollars from butter. I've also made good money from my fat hogs. You have to save your money here just as much as over there and you have to work too, only there's more plowing here and that is tiring.... I like it here very much. Up until now I've had nothing to complain about.

Anton Senger was a German from Russia who went to North Dakota in the spring of 1886. Years later he remembered his introduction to prairie life.

I will never forget my first night on the prairies. The farther we got the bigger the hills were, until, when night came, we were right in the middle of them. I was scared to death, and felt sure some unknown animal would surely eat us up during the night.

We picketed [fenced in] our oxen and rolled in blankets to sleep. But I didn't sleep a wink. There were millions of mosquitoes. Then every little while a coyote would howl on one side then a fox on another, and to make it more miserable for me, a night owl would let out a screech in between. All of those different noises kept the chill running up and down my back all night and I was glad when morning came....

That same winter we spent three days in our sod shack while one of the worst blizzards I ever went through hit the country. We used hay for fuel and that soon gave out. The snow blew so hard we couldn't get out of the house and all we could do was sit inside and try to keep from freezing.

Storms, insects, and drought could all destroy the year's work of farm families. Lillie Marcks went to Kansas with her family from Ohio in 1869. She wrote about a plague of grasshoppers that occurred when she was 12.

I ...looked toward the sun. Grasshoppers by the millions in a solid mass filled the sky. A moving gray-green screen between the sun and earth.

Riding his pony like the wind, father came home telling us more tales of destruction left in the path of the pests. They hit the house, the trees and picket fence. Father said, "Go get your shawls, heavy dresses and quilts. We will cover the cabbage and celery beds. Perhaps we can save that much." Celery was an almost unseen vegetable in that time and place—they wished to save it. They soon were busy spreading garments and coverings of all sorts over the vegetables....

Everyone was excited trying to stop the devastation. Bonfires began to burn thru the garden. "Now I'll get some of

When Eli Yoder needed to build this barn on his Iowa farm, his neighbors arrived to help. Communal "barn raisings" marked the close ties between people in German American communities.

them," Jake [the hired man] said. Picking up a shovel, he ran thru the gate. Along the fence they were piled a foot deep or more, a moving struggling mass. Jake began to dig a trench outside the fence about two feet deep and the width of the shovel. Father gathered sticks and dead leaves. In a few minutes, the ditches were filled with grasshoppers, but they soon saw the fire covered and smothered by grasshoppers. Think of it, grasshoppers putting out a fire.

Ella, my five-year-old sister, was shooing and beating them off the covered garden by means of a long branch someone had given her.... Then all at once, Ella's voice rang out in fear, "I'm on fire!"...I ran to the door and saw a flame going up the back of her dress. In less time than I can tell this, I ran to her and tore off her dress from the shoulders down. Then I turned and looked at the writhing mass of grasshoppers on the garments covering the vegetables, and called, "Ma! Ma! Come here! They are eating up your clothes!"

Sophia Kallenberger Beck was born in Neuburg, in southern Russia, in 1877. She came to the United States with her family in 1886. More than half a century later, in 1939, Beck described her family's first year, near Eureka, South Dakota, where her father made a land claim.

Timbers were hauled from the Missouri river for the roof rafters [of their sod house], over which they laid smaller branches. They covered these with straw and over all they laid sod. The house was about 16 ft. x 22 ft. and had only one room. It had but one door, that leading outside, and two small windows. The bare earth was their floor.... As to furniture, we had none at first. We ate from an old upturned wash tub set on stones. At this table our parents sat on blocks of wood, sawed off portions of a tree trunk. We children sat or sprawled on the floor. We had one kettle, a few plates and cups brought from Russia, and a few spoons. Only one kind of cooked food could, of course, be served at any one meal. We all slept on straw on the floor. Later in the fall and winter, father obtained some lumber and from this made a bed for himself and mother, a table and two benches. Toward spring he also made a bed for us children.

For lighting the room, we had a dish of sheep tallow with a wick of braided cord stuck in it. The cord was first immersed in the melted tallow and allowed to become hard and stiff so it could lie in the dish and one end be twisted upward....

Our Ma had very little time for [a sister, Elsie, who slept in a box and cried a lot] since there was another baby that constantly yelped or whimpered.... I cannot ever remember Ma taking time to hold me.... It seemed so many pioneer mothers had no time, nor patience, nor ability to reason. There was no time to talk or teach, only punish when something went wrong....

Pioneer mothers, like my own, journeyed through life meeting hardships and denials, many dying of a broken heart

A family harvests a crop of sorghum in Wisconsin in 1901. German American farmers typically did not use hired help; instead, the whole family shared in the chores.

Mechanized equipment, like the steam engine used on this South Dakota farm, helped German farmers plant and harvest larger crops.

for want of a word of praise, and few knowing how much they had left behind. These women endured indifference, seldom hearing a word of approval or kindness. What do I remember most? I often found my mother crying, and wonder whether she cried of fatigue, craving a word of recognition, gratitude or praise.... Not only did she look after her household and the hundreds of duties there, she was expected to be first hand help with outdoor work. Her day began at dawn and did not end until the small hours. The family's livelihood often demanded of a woman something beyond human endurance. Cooking and sewing for large families turned women into old ladies before their time.

C. C. Wedel went to Kansas in 1874 from Russia with his family. He recalled the work of his father and mother in the prairie.

They harvested their wheat with scythes they had brought along from Russia, and tied in bundles by hand.... The bound bundles were hauled home, but the horses had only the grass as fodder so they were not strong. Only two wagon loads a day were hauled home under these conditions. As soon as the grain was dry enough, it was rolled out with a big stone. This stone was pulled by a horse and rolled it over the grain threshing it.... This threshed grain was thrown up in the air so the wind could blow the chaff and straw away.... Another name for the [threshing] stone was "Waltzer." The floor or ground where they used this stone had to be very hard so they could scoop up the grain. They also had a *dreschflegel* [flail] made of a long stick with a two by four tied on it and they also used this to beat out the grain.

The women had not forgotten to take some vegetable seeds along from Russia. The seeds did wonderful in the new soil. They planted watermelon seed in sod. The watermelon was unknown here in the United States. Small dwarf peas called Russian Peas also grew and produced seed in the new soil. They had not even forgotten flower seeds....

They started to break the [grass-matted soil of the] prairies either with horses or steers. Those that used steers to plow had

Henry Mueller, a Minnesota farmer, owned this mechanical stump-puller that helped clear the land of timber so that it could be farmed.

The man holding the pitcher of homemade beer is Joseph Erbert II. Around 1886, he arrived in Ellis, Kansas, the site of a colony of Germans from Bukovina (in today's Ukraine). Meeting a later train of immigrants, he saw a young woman step off and said, "That is the girl I will marry." He did, and the picture shows him toasting his wife's family, the Webers.

August Kruger and his wife rake grain on their Wisconsin farm. To Germans who had eked out a living on small plots of overused soil at home, the vast expanse of fertile land in the United States seemed like a dream come true.

problems that required much patience. If the steers saw water puddles, they headed straight to that hole and lay down in there.... The men could do what they wanted but the steers would not budge until they were cooled. They could then be guided back to the field. The settlers broke much land with steers. In this new soil they planted cucumbers, watermelons and corn. They liked the American corn better with its longer ears than the Russian corn; so it was not planted anymore.

One day as I was laying corn [seed] in the furrows which my father had drawn, all of a sudden there came a small herd of antelope running. I had not seen such animals before; so I ran for my life to my father who told me they were not harmful. The settlers also made fire prevention strips by plowing several furrows around their places. Fires were occasionally started by lightning during the hot dry summers. After the fire was over it was of great interest to us boys, where yet a thin column of smoke went, we knew there would be nests of prairie hen eggs. These eggs were baked and we ate them where we found them, were they ever good....

It didn't take too long and each farmer had some hogs which was the winter meat supply. The yearly butchering of the hogs was done with primitive tools from Russia. This was a great day for us children because we could even miss school. In Russia when we had more eggs than we needed, we traded them for either peppernuts or sunflower seeds. We soon found out that eggs could also be traded in the stores here in America for useful articles such as food.

The women were practical in their household efforts. We were always thrilled when they prepared a syrup out of watermelon juice. This syrup was thin, very smooth so that it could not be spread with a knife. Therefore the bread or zwieback was dunked and that was no hindrance for the children. Another kind of syrup was also made, red beet syrup, but that was not as good or well liked since it had a bitter taste to it. After they started to plant sorghum cane, then we made syrup by the barrel full. That was the real stuff; it really tasted good.

At the east end of our village stood the syrup press where the syrup was cooked. People came from near and far to make their syrup. If some of our close friends came, we boys stayed up late even if it got to be midnight. The burned syrup at the bottom of the pan was ours, and was it ever sweet; sweeter than an all day sucker.

Fred Martin came to North Dakota in 1909 with his family. He described the hardships of his early days to his niece.

We were told winters could be pretty rough. Luckily we found a flat-iron cooking and baking stove. We soon realized that on the treeless prairie land there was no fuel to feed that stove for the winter. We sent the boys out to gather *mischt*, dried cow manure, but it was scarce because everybody was gathering it.... We also were introduced to some new fuel possibilities, dried corn cobs, that we had not had in Russia.... One day Ludwig [his brother-in-law] disappeared for hours and turned up with a box full of crude grayish black pieces of something I had never seen before. It turned out to be substance from a top layer of coal. Despite the heavy smoke we realized that that was what we needed for the coming unknown winter. Somehow your Pa saw to it that we had coal piled on the other side of our boxcar home.

The next necessity was food. With twelve of us, we needed a good supply of everything. We sorely needed help, and if it hadn't been for those loaves of bread that good people gave us, things would have gotten unbearable. Our boys had a way of running about and showing their hunger. Some of the more established immigrant families realized our dilemma. One day our oldest son, John, brought me a large hunk of fresh bread soaked in chokecherry syrup, and it was good. The next day a good woman brought us a big kettle of borscht the way we made it in Russia. It was not long before we had several German-from-Russia friends. Even if they were still poor, we felt a longing to be in such a settled position. We thought that if America gave us what they had, we would be thankful and satisfied.

We were two families crowded into a rectangular boxcar for our first winter in America; we were jammed into every inch of space. The bedding we had lugged with us had to be burned because of bedbugs. People rescued us with a gift of fresh corn husks and straw for filling large sacks to serve as mattresses that we placed in one corner where the two sets of parents and three little ones slept. We managed to make a hideaway nook with gunny sack cloth around it. In it stood a five-gallon empty grease can which. . . a farmer had given us as a night toilet, and near it a box of corn husks for toilet paper. We thought that a luxury. Corn and a certain wide-leaved sudan grass were new crops that fascinated me. We had none in Russia. Only better established farmers dared raise corn and sudan grass. What a challenge.

The bandleader and accordion player Lawrence Welk was born near Strasburg, North Dakota, in 1903. His parents were Germans who had emigrated to Russia as children and then moved to the United States in 1892. Welk recalled his boyhood in his 1971 autobiography.

When my parents first arrived there, North Dakota had just been admitted to the Union, and the country around Strasburg, in the south-central part of the state, was still rather wild and harsh and sparsely settled. But to my parents this country offered the kind of wealth for which they had been searching all their lives—it offered freedom and a chance to achieve just as much as their own dreams and initiative inspired them to. Father immediately homesteaded a claim of land about three miles out of Strasburg and began to carve brick-shaped pieces of earth out of the land to build a sod house. It was a crude house, with walls three feet thick, but it served its purpose well, keeping the bitter cold out in the winter and providing a cool haven in the summer.

Germans from Galicia, which became part of Poland after World War I, emigrated to the United States in the 1920s. These families settled near Loma, North Dakota.

ADVENTURES IN THE WEST

Rosa Kleberg went to Texas with her husband in 1834 and settled in what is now Austin County. Conditions were harsh compared to the genteel life they had lived in Germany, but they stuck it out. Later generations of the Kleberg family would become the largest landowners in Texas.

We lived in the house during the rest of the winter. It was very poor, and only in the kitchen was there a fireplace. My father carried on a butcher's trade, while my sister and I took lessons in sewing from a Mrs. Swearingen, and we made clothes for Moore's store. We were all unused to that kind of work, but we felt that we must save our money. When required by necessity, one learns to do what one has never done before.

We had our pleasures, too. Our piano had been much damaged, but I played on it anyway, and the young people of Harrisburg danced to the music. Toward summer, we all took the fever, and it seemed to me as if we would never get rid of it. We had no medicine, and there were of course no doctors....

Circumstances were very different from what we expected. My brothers had pictured pioneer life as one of hunting and fishing, and it was hard for them to settle down to the drudgery and toil of splitting rails and cultivating the field, work which was entirely new to them.

New Braunfels was the first German village in Texas. Prince Carl of Solms-Braunfels, head of a group that had acquired land there, arrived in 1845 to lead the first wagon train to the settlement. The next year, Pastor Louis Ervendberg, the first minister of New Braunfels, arrived. His daughter remembered the early days of the settlement.

Provisions were scarce. We had plenty of fish from the river, but not much meat. Ammunition was too hard to get. Milk and butter, beef and hog meat all came in due time, as did sweet potatoes and a few vegetables but no Irish potatoes for ever so long. We were hungry for fruit and had to be warned by our doctor against eating the fruit of the cactus. Parched barley and dried sweet potatoes were used when coffee gave out, and the men smoked all kinds of weeds when they couldn't get tobacco. Everything that we could not raise for ourselves had to [be] hauled to us overland.

Gilbert J. Jordan grew up on the ranch his grandfather Ernst Jordan had started after emigrating to Texas in the 1840s. Gilbert described some of the ranch work.

Mathilde Hampe aims a stream of cow's milk at a friend at Old Camp Rucker, Arizona. Someone has written on the photo, "From cow to consumer direct."

Shoeing horses (we called it *die Pferde beschlagen*) was a periodic chore. The English word *horseshoes* is really not very descriptive; in German we called them *Hufeisen* ("hoof irons"). Our horses were hard on shoes because they traveled much on coarse sand, gravel, and rocks. Most horses were patient and let us trim their hooves, rasp the bottom flat for an exact fitting, nail on the iron shoe, and rasp the whole hoof smooth, but some horses were natural rebels. It was a tricky business to hammer the flat, sharp nails into the hooves in such a way that the points of the nails emerged at the right place, then to snip off the nails and clinch them. If the nail was driven in too deep, the horse jerked and became lame. Or if the nails were not driven in deep enough, the horseshoe came off before it wore out. There was always a dangerous moment when the nail was driven in and the sharp point emerged. If the horse jerked or jumped, the shoer might get slashed.

There were also regular cleaning jobs to do in the horse yards, the cowpens, the pigpens, the sheepfolds, and the henhouse. This cleanup gave us an abundant supply of manure, and we hauled it as fertilizer to the fields, garden, orchard, and flower beds, just as farmers in Germany do.

Every Saturday we children had to shuck corn by hand and then shell it by the bushel with the corn sheller, which had a crank and a big flywheel. One of us turned the crank until the flywheel hummed, and a second one fed the shucked ears of corn in as fast as the sheller swallowed them.... We saved some of the cobs to be used by the men later [as toilet paper], in place of the Sears-Roebuck catalogues used by the women....

Before the children went to school in the morning and after they returned in the afternoon, they had to help with the regular chores. They fed the milch cows, horses, hogs, chickens, dogs, and cats; they milked the cows, rounded up and counted the sheep in the afternoons, gathered the eggs, chopped and brought in stove wood. The girls worked in the kichen and garden; they helped prepare supper and breakfast; they set and cleared the table, washed the dishes, swept and cleaned the house, and raked the yard from time to time....

R. C. Adelman (with wheelbarrow) and his partner search for silver in Idaho around 1900. John Augustus Sutter, an immigrant from Bavaria, set off the California gold rush of 1849 when gold was found at his mill on the Sacramento River. Though Sutter's business was ruined, many other German immigrants went west in search of precious metals.

The first settlers of New Braunfels, Texas, assemble for a commemorative portrait half a century after their arrival in 1844. The town was named after Prince Karl von Solms-Braunfels, who sponsored the venture.

The children of three German American families in the Yakima Valley of Washington in 1922 take a break from weeding and thinning a field of beets.

Many German immigrants had learned the printing trade at home and founded print shops and newspapers in the United States. This is the office of the New Braunfelser Zeitung (New Braunfels News), one of hundreds of German-language newspapers published in the United States.

We made our own laundry soap out of lard and lye. The mixture had to be boiled until it got stringy, poured into long wooden forms, and then, after it got firm, cut into squares. It turned a dark brown-and-tan color. In my days we had hand-operated washing machines and wringers, but we also used scrub boards and rubbed the work clothes until they were threadbare.

Herman Francis Reinhart was born in 1833 in Saxony, which became part of Prussia. With his parents, two sisters, and a brother he emigrated to the United States in 1841, settling in New York. When Reinhart was 17, he and his brother made their way across the plains to California to seek their fortune in the gold fields. Many years later, Reinhart set down his recollections of the rough times he had experienced. In one of them he described staking a claim with a partner.

About the fourth day of our work on the claim, the weather being very cold, and the water was from the snow in the mountains and as cold as ice, we had only commenced cutting logs for our cabin, and our shoes were in pieces and we could not work long in the ice-cold water, for we had to bail out every time we commenced work and keep bailing while we were at work. On that day a pack train got into Brown [town] with boots and picks, shovels and axes and groceries. The day was cold and blustery but we concluded about noon to go to work and see if we could not make enough to buy us a pair of boots each, at $16 a pair....

So we went to work with our almost naked feet and bailed out the cold water, for the hole would fill up to the level of the creek when we quit work or quit bailing the water out. We worked until it got so cold to our feet that we could not stand it longer, and about two or three o'clock washed up and took our gold over to Barnes' cabin across the creek to weigh it, and found we had made $88 in beautiful heavy coarse gold, good work for two of us in two or three hours.

So we went up to Brown about a half mile above, and bought us each a pair of boots and warm new woolen socks, a sack of flour, a shovel, pick and ax, and some groceries, all for a few hours' work.

Samuel J. Kline was born in Leavenworth, Kansas, in 1859, the son of Jewish immigrants from Munich. Sixty-five years later, he described his family's journey across the plains to Denver in a wagon train.

I recall, from hearsay of course, that the caravan or train consisted of eighty covered wagons, "prairie schooners."... The wagons were large, almost like moving vans one sees these days. They were canvas covered and carried all manner of household and cooking utensils. I remember particularly that ours carried a stove and being cautioned against burning myself....

We had an escort of soldiers. There were very few "settlements" between Leavenworth and Denver, and the Indians were hostile and constantly on the warpath. Every night the wagons were ranged in a wide circle, the animals placed within to prevent stampede or their being run off by the Indians or molested by wild beasts....

Wild animals abounded, particularly coyotes, and buffaloes ranged the plains in countless numbers. Everyone had one or more buffalo robes (hides of buffalo crudely tanned but impervious to cold or moisture). Our fuel consisted of "buffalo chips" (buffalo and cattle dung dried by the sun), which lay everywhere in greatest profusion.

Wilhelm Bürkert came to the United States in 1876. Four years later he was writing home about his adventures in the West.

I went out [to California], as I thought to make my fortune. My savings then amounted to 125 dollars. I and two others bought a piece of land for $300 and started to *minen*. But there was very little gold to be found and also we were dogged by the rough miners and mine owners. Then we were forced to sell our land for $175 after 4 months of hard work. We lost about $35. We took off for Oregon and went on from there eastward, until we came to Colorado. There we worked as guides for the trains of immigrants which move from the eastern states to the West. This job was very dangerous, we had to fight too much with bands of Sioux Indians and with the so-called *"Outlaws"* (that means with white men who break the law). I stayed there about 7 months. Went through the *Indian Territory* [now Oklahoma], where we earned a few talers [German coins] with *trapping*, that means catching beaver, by selling the furs in Texas and from there I came back to the state of Missouri 3 weeks ago.

Mules powered this mill on the Preiss farm in Texas in 1848. It was a Waisenfarm (orphan's farm) founded by a German Lutheran minister to care for children who had lost their parents in an epidemic in 1846.

Three generations of the Lindeman family in their piano factory in New York City in 1910. Germany's musical tradition made instrument making a profitable line of work.

A German American blacksmith in Milwaukee. Typically, he would not only shoe horses but also make iron tools, farm implements, and bands for wagon wheels.

ARTISANS AND LABORERS

Finding a job was the first concern of new immigrants. Nikolaus Schwenck came to the United States in 1854 and finally settled in Chicago. He wrote home to his brother.

After being here three or four weeks without work, I accepted a job, out of boredom, in a smith's workshop, and there I worked for two weeks as a helper. It was in the same shop where Franz Paul [a townsman in Germany] is still working today. They make only carriages and plows there, with 14 fire places in the workshop and about 30 smiths plus the same number of carriage makers, a nice big shop. So things were going quite well with me, but I heard of a place where they were looking for a coppersmith. I went there right away and sure enough, I got the job. And I will stay there for as long as it lasts. My masters are two Americans in partnership, a coppersmith and a tinsmith. We have six coppersmiths in the workshop, among them two Englishmen, three Swedes, and one German, namely my humble self. I like it quite well, only the English language is still a problem, but soon it will surely work out better. I already went to school and tried to force it along, but it just did not go any faster. With regard to the work, it is less than spectacular, we make nothing but pipes which go into locomotives for the railroad. In general, things are very poor in the coppersmith trade. And there can be no question yet about starting my own business. [Schwenck did succeed in opening his own business, which thrived.]

Carl Berthold arrived in America in 1852, at the age of 24. Like many immigrants he found that although jobs paid well, prices were also higher than they had been in Germany. Five years later he wrote his sister and her husband in Germany.

Mississippi is a slave state and I don't want to live here forever. I went to the South because the wages are much better than in the North. But it is *not* so healthy here and the heat is much worse than in the North, but I can take the heat pretty well. I am working here in a *Schapp* [shop] where carriages and harnesses are made. Dear brother-in-law you want to know what I earn here, I work by the piece and make only fine harnesses, my average weekly wage is 15 dollars, which is paid out every Saturday. You may think then that one can get rich in only a year. But that isn't so, because everything is very expensive here, my expenses including board amount to 6 or 7 talers a week.... You should also change your notion of America, you always think when someone comes here from Germany I must have seen him or talked to him, but that isn't so, because things here aren't as close together as Korbach and Lengefeld [two neighboring townships in Germany].

Oscar Ameringer, the son of a woodworker, arrived in the United States in the 1880s. His brother, who had emigrated earlier, helped him find a job in Cincinnati.

Brother...helped me to secure a job in a small furniture factory, and so back to wood shavings and sawdust. There was no resemblance between the work in that furniture factory and father's shop. Here everything was done by machine. Our only task was assembling, gluing together, and finishing, [paid] at so much a chair or table, the two specialties of the factory. Speed came first, quality of workmanship last. So long as the product passed the inspection of the foreman, well and good. How soon it would fall apart was the least of his and our worries.

The work was monotonous, the hours of drudgery ten a day, my wages a dollar a day.

Germans had a virtual monopoly on the brewery business. Alfred Kolb, who worked in a brewery in Chicago at the turn of the 20th century, described his job.

The regular workday in the brewery lasted from 6 to 12 and from 1 to 5 i.e. ten hours. There was no coffee break or late afternoon break. But beer was handed out at 9 and 3 o'clock. Admittedly, everyone drank at other times, also, whenever we got thirsty. As long as you didn't lay your hands on stout or other good brands, the foreman looked in the other direction. By the way, this license was not at all abused. It is true that cheese was preferred at breakfast, "because it makes you so pleasantly thirsty." But I hardly ever saw anyone tipsy; except maybe during overtime, when the men restored their vanishing strength with alcohol. Overtime night work was almost routine; usually until 9, sometimes 10 in the evening. In that case a half-hour break was taken at 6 o'clock, which would have been very nice if the lunch break had not been shortened accordingly. Thus the total length of work was from 14 to 15 hours. In addition, we worked on Sunday morning from 6 to 12 o'clock.

Many Germans went into the bakery business, both as owners and workers. In 1881 the German American newspaper Chicagoer Arbeiter-Zeitung [Workers' News] *described conditions in that trade.*

The average workweek for the bakery hands in Chicago amount to six days, the average workday to approximately 15 3/4 hours.

This workday, however, is stretched to 17 or 18 hours

Brewery workers in Cincinnati around 1870. Pennsylvania Germans founded the first German brewery in Baltimore in 1748. By the end of the 19th century, there were hundreds, each with its own formula for making beer. The passage of the 18th amendment in 1919, which made alcoholic beverages illegal, wiped them out.

due to various short interruptions spread throughout the day. The two to three hours between the punching of the dough and its rising are not taken into account in the above-mentioned 15 3/4 hours....

The journeymen bakers' sleeping quarters are almost never whitewashed. The sleeping quarters in the bakeries are never cleaned, scrubbed, or even free of bedbugs and other vermin, like rats, mice, etc., unless, that is, the bakery boss—not for the sake of the bakery hands, but rather to prevent his beams from completely rotting away—undertakes the most urgent, long overdue repairs to the house....

The food placed in front of the hands in the morning and at night is unappetizing and almost always consists of warmed-up leftovers from the last midday meal. 80% of all the journeymen bakers complain about their board. Especially in the morning, exhausted after their 16-hour day, the hands have to wait several hours for their "breakfast" or go to bed hungry.

Ernest L. Meyer was born in Milwaukee in the late 19th century. His father was an editor for Germania, *one of the city's many German-language newspapers. Young Ernest enjoyed visiting the office. One of its busiest employees, especially on hot summer days, was a man called Heinz, who was the* Kesseljunge, *or "bucket boy." Meyer later described him.*

The bucket boy was not a boy at all. He [had been] a superannuated bookkeeper in a brewery. But at sixty he was let go because his eyesight failed.

Yet Heinz was lucky to get the job of *Kesseljunge* at the *Germania,* for it was an important post and his services were in great demand. He himself constructed the tools of his trade; two poles, each about five feet long and each artfully and deeply notched. The notches were designed firmly to hold the handles of one-quart beer pails, and Heinz had mastered the art of carrying six full buckets on each pole without spilling a drop....

Heinz and I became friends, and he took such a fancy to me that he told me some of the secrets of his trade....

"There is one important thing to remember," he said, "and that is, never take the passenger elevator even if you are in haste and the thirst of the gentlemen is great. For sometimes the passenger elevator is a bit crowded, and people jostle against me, and though I am careful as can be it happens that I may spill a little beer out of one of the twelve cans....

"So that is why I always take the freight elevator in the back, even though it is a slow affair and it is a greater distance from this elevator to the saloon at the corner. This, too, has an added inconvenience because beer has a bad habit of shrinking in transit when the foam goes down in the can. That is why, when I enter the basement and before I ring the bell for the freight elevator, I give each can a good *Schwupptich.*"

I asked Heinz what a *Schwupptich* was, and he explained patiently:

A bucket boy carrying pails of beer to the employees of an iron-works in Milwaukee. Employees demanded such a service to make the workday easier.

"I take each can and twirl it a bit, round and round, but slowly, in this fashion. That is a *Schwupptich*. This causes a certain agitation in the beer, making the foam rise, and consequently again filling my buckets. Therefore a *Schwupptich* is highly essential, otherwise the gentlemen, seeing the buckets only partially filled, might suspect that I had taken a private sip in the basement."

"Do you, Heinz?"

Heinz looked at me, astonished.

"But of course I do," he cried.

Ludwig Hofmeister came to the United States in 1925. His family had worked in the Black Forest as clock and watch makers. He told an interviewer about his first job in the United States.

I was home quite some time, until my brother got me a job in a jewelry store. This man had a big clock business, and I was good already in clocks. I could fix clocks and watches, and whatever I didn't know he showed me. He spoke German. I worked there until 1929 or 1930. That's when things really got bad, and then he couldn't keep me any longer....

When the Depression came, that was really bad for watchmaking—for the jewelry stores and for watch repairs. People couldn't pay the prices. It was impossible. I tell you what happened. People working for the trade, for jewelers, didn't have enough work anymore. They were starving. So they figured, "Well, why don't we go out on the street and give the same price that we worked for the trade and give it to the people?" Then they opened up a cut-rate watch company. You probably never heard of this. This was a very unusual thing. They repaired watches for a dollar and a half! That was including mainsprings and I don't know what else. It wasn't exactly the greatest job anymore. You had to work fast and things like that. But in the meantime, they did business. The Depression helped them. I worked there for quite some time. It wasn't exactly what I liked but was better than nothing. I had to change a great deal, but I adapted myself somehow. It wasn't that hard on me.

Then I got the idea, why don't I do this on my own? Why should I work here and slave away like crazy? Why don't I do this out on my own and work for the trade or open up a place? My brother came in and we did it. This was in 1937. I rented an office downtown, and I went out to different towns and picked up work, and we made a living. In the beginning we didn't do so well, but eventually it worked out pretty good. Then we went down on the street and had a store on the street, on the ground floor, like a retail store. We did very well there, too. I mean we made a living—you know what I mean? We made as much as we made when we were working for somebody else, and that was good enough for us.

A carriage and wagon shop in Milwaukee. Small craftsworkers' shops like this gradually disappeared in the 20th century, driven out of business by factories that produced cheaper products on a mass scale.

The Lititz Springs Pretzel Company was owned by Lewis C. Haines (background), who is unloading a tray of pretzels that has come up from the baking room below. His son Rob weighs them and packs them into cans. Lititz, Pennsylvania, was the first town in the United States where pretzels were made.

FACTORIES AND MINES

Some of the first German immigrants were weavers. Their trade was then a skilled craft, done by hand. By the time 31-year-old Martin Weitz arrived from Germany in 1854, however, mechanical looms in factories had driven skilled weavers like him out of business. Weitz wrote his father a letter describing the work.

On March 16th in the newspaper there was a call for 25 weavers in a wool *Fektori* [factory] in Rockville in the state of Connecticut to sign up on the 17th at 6 o'clock in the evening. We went there and were accepted. On the 18th we got on the steamboat and went to Hartford, from there to Rockville, first I had to sell my watch otherwise I couldn't have gotten there. When we arrived there in the afternoon they said you have to work nights from 6:30 in the evening till 6:30 in the morning, then we were shocked. I said I didn't care [as long as] I only have work. The looms all work by themselves, they are all driven by water, I'd never woven on such a loom. I went to a fellow who taught me during the day, then work started. It didn't go well of course in the beginning, they do difficult patterns. In March and April I didn't earn much, I hardly had enough for Board, in July 19 1/2 dollars. Now it's getting better, if you do good work, you earn 18-20 to 24-25 dollars.

In 1855, Peter Klein emigrated from Güchenbach, in the Saar Valley, where coal mining was an important industry. He found work in a mine in Pottsville, Pennsylvania, and wrote to his parents.

You write in your last letter that you want to know how coal is worked here. I can tell you that most of the work is done with [blasting] powder and the shifts

Meat trimmers in the Chicago Stockyards. German butchers were among the skilled laborers who received relatively higher wages in the United States than in Germany.

Metalworkers in the McCormick Reaper Works in Chicago. In 1886, German American workers in this factory organized a strike, demanding an eight-hour workday. The company hired non-union members as replacements, and clashes between union members and police officers resulted.

are 10 hours, and the piece rates are like in Germany, and one man digs almost as much coal as six in Germany.

Then you should know that they have coal here that is from 4 feet up to 20 and 30 and 40 feet high. That's to let you know there's only a handful of coal in Germany, compared to here.

Women and children worked to supplement the family income. Often they were exploited and sometimes not even paid, as can be seen in the following account by a young girl that appeared in the Chicagoer Arbeiter-Zeitung *of December 23, 1884. She worked in the Bonde & Co. Candy Factory.*

A young German American steelworker. Immigrants with skills could find good jobs in the steel industry. Companies provided housing and encouraged workers to marry so that the work force would be stable.

When we started several weeks ago, Mr. Bonde promised us $2.50 a week. But shortly after wards he came and said that we had to do piece-work. Then he set the prices so that even if we worked as hard as we could, we could barely earn $1.50. We complained and got a raise. We were then in a position to earn $2.00-$2.75 a week. But we didn't receive the money we had earned. Every week 40 cents to $1.00 were deducted so that we only got our full pay, but not the money that had been held back earlier. The next Monday we all went to Mr. Bonde and asked him to tell us if and when he intended to pay us that money, otherwise we were going to walk out. "You cheeky little kid! You've stirred up all my girls. Get out of this shop on the double! and the rest of you get back to your work." But my friends didn't go back to work.

A letter to the Chicagoer Arbeiter-Zeitung *signed "a sewing girl" described the conditions for people in her trade.*

About 300 comrades of mine are employed in the corset factory at 250-252 Wabash Avenue, and we truly have to work for starvation wages. But this isn't the worst of it. Take a look at the following list of fines set for the girls, and tell me whether this robbery—an allegedly nonviolent and legal process—isn't just as bad as those things described by your police reporter:

Breaking a needle (at 1/4 ¢ ea.) or a
    stay (at 1/2 ¢ ea.) .... 5¢
Leaving the machine .... 5¢
For each drop of oil on the work .... 5¢
For cleaning the work of oil spots .... 20¢
Slightly late in the morning .... 5¢
Slightly late at midday (for ex., 31 minutes
    instead of 30) .... 5¢
Breaking a "bobbin" .... .5¢
For not coming to work, in addition to the
    loss of wages .... 20¢

The highest price paid in the factory for a dozen corsets is 30 to 50¢. Now just imagine what is left over for the poor girls after all the fines are deducted, especially with our "foremen," Jim Stone and Grace Murphy, running around

Women at the H. J. Heinz factory pack pickles into jars. Henry John Heinz, the son of a German immigrant, believed in treating his employees humanely. He provided them with free medical care, homemaking classes, a swimming pool, a gymnasium, and weekly manicures.

The Haymarket Square Riot

Early in 1886, the Knights of Labor, then the country's leading labor organization, started a campaign to win an eight-hour day for workers. On May 1, 40,000 workers in Chicago went out on strike. Two days later, police attacked a group of strikers at the McCormick Reaper plant, fatally shooting two unarmed men. August Spies, editor of the German-language newspaper *Arbeiter-Zeitung,* called for a protest rally at Haymarket Square the following evening.

The rally began peacefully. Because it was raining, the crowd numbered only about 3,000 people. Chicago's mayor listened to some of the speakers and left, thinking there would be no trouble. But an anti-labor police captain decided to break up the rally. When he led a group of officers into the square, a bomb went off. The police began to fire into the crowd; some demonstrators fired back. Seven policemen and four demonstrators were killed.

Eight well-known radicals, including Spies, were charged as "accessories" to the murder of a policeman killed by the bomb. The actual bomb thrower was never discovered.

The judge at the trial was openly hostile to the defendants. Despite flimsy evidence, the eight men were convicted. Seven were sentenced to hang. Later, two had their sentences commuted to life imprisonment. One committed suicide before his execution. The other four were hanged in November 1887.

Newspapers around the country used the Haymarket riot to condemn the Knights of Labor, and its membership declined drastically. Factory owners branded all strikers as dreaded anarchists.

In 1893, German-born John Peter Altgeld was elected governor of Illinois. He pardoned the three remaining prisoners because they had not received a fair trial. The resulting uproar ended his political career.

arbitrarily imposing fines. Mr. Gustav Florsheim and D. H. Bal, members of the firm, endorse everything they say.

Besides which the water is almost undrinkable and sewer gases fill the house.

Alfred Kolb, who arrived in Chicago around the turn of the 20th century, found a job in a bicycle factory. He recalled the working conditions.

The assembly room was a huge hall twenty-seven windows long, its entire length traversed by wide work benches full of vises, tools, and miscellaneous equipment. The finished bicycle parts were carried here in a continuous stream on a rattling conveyor belt and then fitted together to make whole bicycles. The principle of the division of labor was thereby applied extensively, and the human hand was replaced by a machine wherever possible. Even the turning of the screws was done mechanically by devises similar to that thin, whirring tube whose painful acquaintance we make when a tooth is filled.

The foreman led me to one of the long benches and instructed me in the work. It required neither skill nor much exertion, consisting of a few constantly repeated hand movements on the front wheels of the bicycles. The man to my left prepared the axles. I stuck them through the hubs, and after I had checked the ball bearings, pulled a greased felt washer over it. The neighbor to my right put on nuts. The next ground down the ends of the spokes, slipped the rubber tires over, inflated them until they were firm, and so forth, until the wheel all ready to go, arrived at the end of the table, where it was then inserted into the front fork of the frame....

The workday in the bicycle factories lasted the ten hours typical of the city; in the beginning there were also up to three hours overtime in the evening, but they were short-lived. The entire operation of the plants was cut into seasonal work. At the beginning of winter, one worked day and night; they had already stopped when I began there. As summer approached, more than three-quarters of all the workers were gradually let go or else stayed away of their own accord because they found a more appealing job somewhere else. And those of us who remained still only got five days a week until finally that stopped, too, and the business shut down for weeks on end. But as far as I know, the daily wages were not reduced.

I repeatedly heard it expressed, and mainly from the mouths of newly arrived immigrants, that work was not so strenuous back home in Germany as here in America. There may be some truth to that; in America, in any case, one never stops hearing the warning, "Get going! Hurry up!" But no rules without exceptions. At first I had worked with all my might so as not to fall behind my comrades. Until one of them asked scornfully, "Hey, are you crazy? Working just like you're on piecework? Don't be a fool!" And when I answered that I didn't want to promote my dismissal, he laughed at me.

"That's no reason to run yourself into the ground, greenhorn. Look at me: am I knocking myself out? The main thing is to get in good with the foreman. His father runs a saloon. Whoever goes there and spends a bit doesn't have to worry about being thrown out."

This last tip was not completely unfounded. Despite the unrelenting pressure, there were several loafers among us who dawdled away the time without being punished for it, trusting in such behind the scene influences. The worst of the lot was a Bavarian, a true genius at doing nothing, and who on top of it all worked himself into a rage over the heavy labor. He worked near me, and I can still see him today, short and fat, alternately taking snuff and grumblin: "Holy Jesus, what a treadmill. Don't it never end? Holy Mother of Altötting [a Bavarian religious shrine]!"

The work began at seven in the morning and lasted until 5:30 in the evening. The half-hour lunch break was at noon. Sitting on the work benches, the married workers then ate the snack they had brought from home. We single workers hurried to the nearby cheap kitchens or saloons. At noon, as well as during overtime at night, we were allowed to get beer, for which most kept special tin containers. A half liter cost five cents. It was fetched by the errand boys working in the hall, who on payday got a few extra cents for it. Aside from these times, the consumption of alcohol was not permitted and the rule was strictly observed. Whoever got thirsty during work drank from the bad tap water.

A German American woman works in a chocolate candy factory. Germans were in demand for food-related industries because they were considered cleaner than other immigrant groups.

Hatmaking was another skilled trade in which Germans excelled. This factory is in Milwaukee.

WOMEN'S WORK

Using women to serve customers added to the family atmosphere that most German restaurants and beerhalls maintained. This waitress worked at a beer-garden in Cincinnati.

In the 1880s, Helen Campbell, a 19th-century social worker, interviewed a woman named Lisa, "a little over 50 years old, whose husband is a painter." They had left Germany in the 1870s to join their oldest daughter, who had emigrated earlier. At first, the husband found work and the family prospered. But he became ill because of lead in the paint he worked with. To earn money, Lisa and her daughter Gretchen took in piecework. Sewing pieces of cloth into jackets for about 30 cents apiece, they eked out a living. As Lisa described the work, Campbell wrote down what she said, trying to reproduce her accented English.

I know not if we shall live at all. For see. We two, my Gretchen and I, we make but ten [jackets] for a day. Tree [three] dollar?... It is early that we begin—seven, maybe—and all day we shall sew and sew. We eat no warm essen [food]. On table dere is bread and beer in pitcher and cheese to-day. We sit not down [to eat], for time goes away so. No, we stand and eat as we must, and sew more and more. Ten jackets to one day—so Gretchen and me can make; ten jackets to one day, but we sit always—we not go out. It is fourteen hours efery day—yes; many time sixteen—we work and work. Then we fall on bed and sleep, and when we wake again it is work always. And I must stop a leetle; not much, but a leetle, for my back have such pain that I fall on the bed to say, "Ach Gott!" Is it living to work so in this rich, free America?... It is because America is best that we come, but how is it best to die because it is always work and no joy, no hope, never one so small stop.

Many German women became domestic workers. Agnes M., a young woman who lived in New York, wrote about such work in the magazine Independent *in 1903.*

I was an apprentice in a Sixth Avenue millinery store earning $4 a week. I only paid $3 a week for board, and was soon earning extra money by making dresses and hats at home for customers of my own, so that it was a great change from Germany. But the hours in the millinery store were the same as in Germany, and there was overtime, too, occasionally; and though I was not paid for it I felt that I wanted something different—more time to myself and a different way of living. I wanted more pleasure. Our house was dull, and though I went to Coney Island or to Harlem picnic park with the other girls now and then, I thought I'd like a change.

So I went out to service, getting $22 a month as a nursery governess in a family where there were three servants besides the cook.

A picnic attended by the office staff of the Miller Brewing Company. Women worked as secretaries, accountants, and clerks in conditions that were preferable to those associated with most other occupations open to women at the time.

I had three children to attend to, one four, one six and one seven years of age. The one who was six years of age was a boy; the other two were girls. I had to look after them, to play with them, to take them about and amuse them, and to teach them German—which was easy to me, because I knew so little English. They were the children of a German mother, who talked to them in their own language, so they already knew something of it. I got along with these children very well and stayed with them for two years, teaching them what I knew and going out to a picnic or a ball or something of that sort about once a week, for I am very fond of dancing.

We went to Newport and took a cottage there in the summer time, and our house was full of company.... I enjoyed life with this family and they seemed to like me, for they kept me till the children were ready to go to school. After I left them I went into another family, where there were a very old man and his son and granddaughter who was married and had two children. They had a house up on Riverside Drive, and the old man was very rich. The house was splendid and they had five carriages and ten horses, and a pair of Shetland ponies for the children. There were twelve servants, and I dined with the housekeeper and butler, of course—because we had to draw the line. I got $25 a month here and two afternoons a week, and if I wanted to go any place in particular they let me off for it....

Wherever I have been employed here the food has always been excellent; in fact, precisely the same as that furnished to the employer's family. In Germany it is not so. Servants are all put on allowances, and their food is very different from that given to their masters.

Anna Groener came to the United States from Germany in 1910 with her aunt and uncle. They ran a saloon and bought some apartment houses. After four years in school, Anna worked scrubbing rooms for her relatives. Then, as she remembered, she started to work on her own.

My first job outside the home was with Bamberger's as a waitress. I worked there from the age of eighteen until I was twenty-three. It was in downtown Newark, and I took a trolley car to work. I worked from eleven to five and got twelve dollars a week. Out of the twelve dollars, I paid seven to my aunt for room and board.

We weren't allowed to have tips. We asked for a raise, so they gave us a half cent on every dollar that we took in. If you didn't drop anything, you got a fifty-cents bonus a week. That bonus I put in a little jar and saved.

I loved this job. The bosses were nice, and we worked with a lot of other ethnic groups. There were a lot of blacks in the kitchen, and some Polish and American girls. I made friends with some of them too. They all liked me.

These women worked in the mending room of the Ayer textile mill in Lawrence, Massachusetts. Anna Lutz, front row at left, was one of many German Americans employed there.

The National Council of Jewish Women conducts a welcoming ceremony for refugees from Nazi Germany. The NCJW, established in 1893, also provided job training and other instruction to help immigrants adujst to American life.

Walter Reuther

On May 26, 1937, Walter Reuther and three other members of the United Automobile Workers union (UAW) met at a footbridge leading to the Ford factory in Detroit. Reuther and his colleagues began to distribute pro-union leaflets.

Earlier in the decade, the UAW had won recognition from other major auto manufacturers, but Henry Ford stubbornly resisted unionization of his factories. Forty men from Ford's security force surrounded Reuther and the other organizers.

Reuther described what happened: "The men...picked me up about eight different times and threw me down on my back on the concrete...kicked me in the face, head, and other parts of my body...and threw me down the first flight of stairs. I lay there and they picked me up and began to kick me down the total flight of steps." Photographs of the vicious beating appeared in newspapers, winning public support for the union. Reuther became a hero to the nation's workers.

The labor movement had been part of Reuther's life since childhood. His father, Valentine Reuther, who had emigrated from Germany in 1892, helped found a union of brewery workers in Wheeling, West Virginia. Val Reuther read books on politics, economics, and socialism to young Walter and his brothers, Roy and Victor, who would also become UAW organizers.

In 1927 Walter Reuther, aged 20, went to Detroit, where he worked in a Ford factory for five years. After giving speeches for the socialist candidate for President, he was fired. Walter and his brother Victor left for Europe, where they visited their father's home village and saw the ominous effects of Hitler's Nazi regime. They traveled to the Soviet Union, where they worked in an auto factory before returning to the United States in 1935.

After Reuther became president of the UAW in 1946, he worked to rid the union of communist influence. Under his leadership, the UAW helped to ease racial tensions in the industry. Reuther was one of the organizers of the great civil rights rally in Washington in 1963. His tragic death in a plane crash in 1970 cost the labor movement one of its most skillful and respected leaders.

THE UNION MOVEMENT

Many German Americans were active in the union movement. In the 1880s, while working in a furniture factory in Cincinnati, Oscar Ameringer joined a union that was part of the Knights of Labor. The Knights were a militant group that aimed to organize all workers—unskilled and skilled, blacks and whites—into one powerful nationwide organization. Some of its members were radicals who advocated socialist revolution, and a few were anarchists who opposed all forms of government. Ameringer recalled his union days.

In order to become a member of the Knights I was compelled to add two to my almost sixteen years. [Members had to be 18.] But whatever I lacked in age I more than made up in enthusiasm for the cause of less work and more pay. The organization I had joined was a branch of the *Deutsche Holz Arbeiter Verein*—German wood-workers' union—affiliated with the Knights of Labor.... The membership was almost exclusively German and seasoned with a good sprinkling of anarchists. Prior to the first of May, 1886, when the eight-hour-day strike was to be launched, there had been groups of older or more militant members manufacturing bombs out of gas pipes. All of us expected violence, I suppose.

Too young to be admitted to the inner circle, I had converted a wood rasp [tool] into a dagger, in anticipation of the revolution just around the corner. The prelude to the revolution was the May Day parade in which I marched, bloody upheaval in heart and dagger beneath my coat tail. Only red flags were carried in that first May Day parade, and the only song we sang was the "Arbeiters Marseillaise," the battle cry of the rising proletariat [working class]. Even the May Day edition of the *Arbeiter Zeitung [Workers' News]* was printed on red paper. Testifying further to the revolutionary intent of the occasion, a workers' battalion of four hundred Springfield rifles headed the procession. It was the *Lehr und Wehr Verein,* the educational and protective society of embattled toil.

Unfortunately for the pending revolution, the forces of law and order in the city made no attempt to interfere. Whether plutocracy [government of the rich] had already abdicated, or, considering that it takes two to make a fight, had taken the wiser course, I never discovered. And so we just marched and marched and sang and sang, until with burning feet and parched throats we distributed our forces among the saloons along the line of march where we celebrated the first victory of the eight-hour movement with beer, free lunch, and pinochle.

German men were important in the union movement in the Midwest; women played a less active role. A letter to the Chicagoer Arbeiter-Zeitung *from a Mrs. M. B. explains why.*

As a member of the Women's Mutual Benefit Society, Lassalle No. 1, I've often heard it asked: Why are there so few working women in these kinds of societies? You don't have to go too far to find the answer: quite simply, they're not allowed to join, their husbands won't have it! Hard to believe and yet true. A man, a worker, himself, a member of a union or society, which indirectly as well as directly serve the same cause, prevents his wife from taking part. I myself know women who would think rationally enough to decide to dedicate themselves to such a noble cause: but women must first get permission from their Herr Husbands. Her mate, however, knows how to convince her how indecorous it would be for a woman to attend a meeting, tells her that she really belongs in the kitchen, doing the housework, etc.; in a word, nothing can come of it. And this is not right! Each worker whose thoughts are in the right place should encourage his wife to join such a society, for women's societies can only achieve something of magnitude if they become strong. And that's why you who read this should shake yourselves out of your lassitude; you've waited too long as it is. A city like Chicago should show other cities in the union what women together can achieve.

As Others Saw Them

German Americans were among the leaders of the labor unions that sprang up in the 1870s. One of these was the Eight Hour League, which fought for a regular workday of eight hours. Its leaders called on workers to assemble in a march of solidarity in New York City in June 1872. A reporter for the New York Sun *described the march:*

The principal thoroughfares of the city rang with cheers yesterday. The workingmen's parade attracted crowds of spectators in the Bowery, in Broadway, in Fourteenth street and Twenty-third street, and all the other streets through which the procession passed. The houses in the Bowery near where the procession was formed displayed flags and the whole street was gay with the colors of the United States and Germany. As the majority of the workingmen in the parade are Germans, so the greater part of the spectators seemed to be of Teutonic [German] origin. It was a gala day for the workingmen of New York. Thousands who did not march in the procession stood on the sidewalks and attested the interest they felt in the parade by enthusiastic cheering.

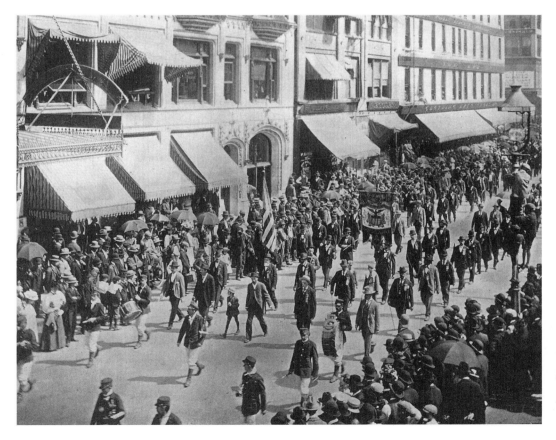

A Labor Day parade in Chicago in 1897. The group shown in the picture is the Carpenters and Joiners Union, which had a significant German American membership.

The Dick family in their backyard arbor in Milwaukee. In summer, the vines growing on the open roof provided shade.

CHAPTER FIVE

PUTTING DOWN ROOTS

The first German immigrants founded their own community—Germantown, Pennsylvania—setting a pattern for the millions of Germans who followed. Until the 20th century, German Americans preserved their language and culture by settling with others who shared a German heritage. The map of the United States is dotted with German names where the immigrants put down roots: New Braunfels, Texas; New Brunswick, New Jersey; New Berlin, Wisconsin; Rhinebeck, New York; Bismarck, North Dakota.

Similarly, Germans who settled in major cities congregated in their own neighborhoods. Shop signs and the spoken language remained German in the *Kleindeutschlands* (little Germanies) of New York, Chicago, Cincinnati, Milwaukee, St. Louis, and Baltimore until well into the 20th century. In 1914, Cincinnati had four daily German newspapers, four hospitals staffed by German-speaking doctors and nurses, and more than 70 churches where the services were in German.

Preservation of the German language was the key to maintaining the cultural traditions that united the German American communities. Not only were church services in German, but so were classes in

the public schools of such cities as Cincinnati, Milwaukee, and St. Louis. At the beginning of World War I, more than 500 German-language newspapers were being published in the United States.

Within the tightly knit German American neighborhoods and communities, family ties were the strongest bonds. In 1883, *Fackel,* the Sunday edition of the *Chicagoer Arbeiter-Zeitung,* asserted that "the man is the head of the family, its protector, its representative outside the home; the woman is the soul of the family, its guardian angel, its inner compass." Mathilde Anneke, who emigrated to Milwaukee in the 1850s, was an active member of the women's rights movement, publisher of the *Deutsche Frauen-Zeitung (German Wives' News)*— and was an exception. Most German American women accepted their traditional roles, expressed as *Kirche, Kinder, Küche* (church, children, kitchen).

To some German Americans, maintaining their heritage was a religious duty. The Mennonites who first arrived in 1683 opposed the taking of oaths and resorting to violence of any kind. The men wore black clothing and hats; the women covered their heads in public and wore long, plain dresses. Because buttons were luxuries of the rich in 17th-century Germany, the Menno-

nites did not use them. Avoiding what they saw as corrupting influences of the modern world, they preferred to live in their own communities, which spread from Pennsylvania to Nebraska and Kansas. Today some Mennonites have adapted to modern ways, but the most conservative group, the Amish, still do not use electricity, automobiles, or motorized farm equipment. The Hutterites, a similar group, also maintain the German language and traditional customs in their communities.

Of course, the vast majority of German Americans were not Mennonites. Lutherans, members of the Reformed church, Methodists, Roman Catholics, and Jews were among the millions who arrived over the past three centuries. Until the 20th century, however, most of them still clung to their German American identity.

The Missouri Synod (governing council) of the Lutheran church, for example, was founded in 1847 by German Lutherans who had left Saxony. Similarly, the German Methodist church was a separate branch of the American Methodist Episcopal church until 1924.

German American Catholics made up about one-third of all American Catholics in the 1890s. Their preference for German-speaking priests created conflict with the bishops of the American Catholic church, most of whom were Irish. When German American priests in several cities petitioned the church hierarchy in Rome for equal treatment, an Irish American bishop in Louisville declared, "If these German prelates are allowed special legislation as Germans...we will be looked upon as a German church in an English-speaking country." The controversy eventually died down with the creation of separate parishes for German Americans as well as for Poles, Italians, and other non-English-speaking Catholics.

After 1830, Jews from the various states of Germany began to arrive in large numbers. They soon formed their own congregations separate from those founded earlier by the descendants of Spanish Jews. German-trained rabbis such as Isaac M. Wise from Bohemia introduced the ideas of Reform Judaism to the United States. Nineteenth-century German American Jews tended to participate in the social and intellectual life of the larger German American community. They were typically as proud of their German cultural heritage as of their Jewish religious identity. Toward the end of the 19th century, they used their national heritage to distinguish themselves from less prosperous Russian Jewish immigrants whom they considered socially inferior.

Wherever they settled, German Americans organized their own associations and clubs. One of the earliest, the Sons of Hermann, was founded in New York in 1840 to foster German customs and language and to aid financially needy members. By the turn of the century it had branches in many other cities. (Hermann, the organization's namesake, was a Germanic folk hero whose men defeated three Roman legions at the Battle of Teutoburg Forest in the year A.D. 9.) There were many other German American mutual aid societies, which provided life insurance, medical care, and jobless benefits for members.

Countless social clubs, called *Vereine* (the singular is *Verein),* were formed in communities large

Women at the South St. Louis Turnverein *prepare a bar drill that they will demonstrate at the city's 1904 World's Fair.*

and small. As Carl Entenmann told the Historical Association of Los Angeles in 1929, "We have a saying that when three Germans meet they start a Society." Some *Vereine* were associations of people from the same state in Germany, such as the Schwaben Verein. They organized social and cultural activities and sometimes took part in politics.

Other *Vereine* met for a specific purpose, such as the *Turnvereine* or gymnastic clubs, called turner societies in English. Part of a physical-culture movement founded in 1811 in Germany, the *Turnvereine* sought to promote health through exercise and gymnastics. The movement also had a socialist bent. The first American *Turnverein* was established by Friedrich Hecker, one of the " '48ers" who had played an important role in the failed revolutions in Germany. Spreading to virtually every large German American community in the 1850s, the *Turnverein* movement also carried out military drills. In the Civil War they formed militias to fight on the Union side.

Equally popular were *Gesangvereine* and *Sängerbund,* German singing societies. Choral singing was a beloved tradition of long standing in Germany. The first American *Gesangverein,* founded in Philadelphia in 1835, was soon followed by others in Baltimore, New York, Chicago, Cincinnati, and Charleston. The *Gesangvereine* organized *Sängerfests* (singing festivals), often in May and October, which featured a mix of German folk songs and classical music.

Many *Vereine* met in neighborhood German beer halls, which were not the same as what we now call bars. The spirit in the beer halls was marked by *gemütlichkeit,* or "good fellowship." Families came there to enjoy the food, song, and socializing. Orchestras played German music, and the walls were decorated with paintings of scenes in Germany. Many kinds of German *wurst* (sausage), *schnitzel* (veal cutlet), and *sauerkraut* were on the menu. In summer, the crowd moved outdoors to an enclosed garden.

By the beginning of the 20th century, most German Americans felt that their place in U.S. society was secure. The German-American Alliance, founded in 1901, claimed 3 million members by 1916. It encouraged the continued use of German in public schools, opposed limits on new immigration, and fought against the movement to prohibit the sale of alcoholic beverages in the United States.

However, the outbreak of World War I in Europe in 1914 brought an abrupt change in German American life. In August, Germany sent troops across the neutral nation of Belgium to attack France. Britain entered the war on France's side.

Some German Americans defended Germany's war policy, but most merely urged the United States not to get involved in the war. Other Americans generally agreed, but U.S. banks made loans to England and France, allowing them to buy billions of dollars' worth of war matériel from U.S. companies.

In 1915 a German submarine sank the British ocean liner *Lusitania,* causing the deaths of more than 1,000 civilian passengers, including 128 Americans. When the United States protested vigorously, the German government promised to modify its policy of unrestricted submarine attacks on merchant ships.

The U.S. President, Woodrow Wilson, ran for reelection in 1916 with the slogan "He kept us out of war." But after Wilson's victory the German government resumed unrestricted submarine warfare. Furthermore, it was revealed that Germany had attempted to persuade Mexico to attack the United States if it entered the war. Ultimately, on April 6, 1917, the United States declared war on Germany.

Throughout the United States, all German Americans now came under suspicion of disloyalty. During the years 1917 and 1918 many German Americans were

In 1909, the Ziegler family celebrates the Fourth of July in Junction City, Kansas.

jailed for speaking out or writing in opposition to American involvement in the war. German businesses and homes were vandalized, and "patriotic" mobs sometimes attacked German American citizens. Robert Prager, an outspoken immigrant from Dresden, was lynched in Illinois in 1918.

The home-front battle against all things Germanic went to ridiculous lengths. Symphony orchestras were banned from playing German music, and German books were publicly burned, even in such bastions of German American life as Cincinnati. Streets, towns, and even foods were given new, non-German names. The frankfurter became the hot dog, sauerkraut became liberty cabbage, and German shepherd dogs were now called Alsatians.

After the war, anti-German prejudice continued. The use of the German language in schools and churches sharply declined. Many German American clubs disbanded, and newspapers ceased publication. Never again would the German American community be as strong and vital as before 1917. Herbert Hoover, who in 1929 became the first U.S. President of German descent, did not publicize his roots.

In the 1930s, Adolf Hitler came to power in Germany. The *Deutschamerikanische Volksbund* (German American People's League) was formed in 1936 to support Hitler's Nazi government. The Bund, as it was called, attracted attention with public rallies at which the Nazi swastika was displayed. However, its membership never exceeded 25,000 people, and most German Americans were unsympathetic to Hitler's Nazi doctrine.

During World War II, German Americans did not encounter the accusations of disloyalty they had faced earlier. In fact, the commander-in-chief of the Allied military forces that defeated Nazi Germany was Dwight David Eisenhower, another German American.

A membership certificate in the Order of the Sons of Hermann. At the top are scenes from the Battle of Teutoberg Forest in the year A.D. 9, in which Hermann led Germans to victory over soldiers of the Roman Empire. The motto of the Order, between the two seals at bottom, translates as "Friendship, Charity, and Loyalty."

No German American community was complete without a band to provide lively music at festive occasions. This is Buch's Brass Band of Lawrence, Kansas, around 1880.

COMMUNITIES

German communities were enriched with many Vereine—*clubs and associations. In his autobiography, Carl Schurz, who came to the United States in 1852, described how the "'48ers" established* Vereine *in Milwaukee.*

They at once proceeded to enliven society with artistic enterprises. One of their first and most important achievements was the organization of the "Musical Society" of Milwaukee, which, in an amazingly short time, was able to produce oratorios and light operas in a really creditable manner. The "German Turn Verein" not only cultivated the gymnastic arts for the benefit of its own members, but it produced "living pictures" and similar exhibitions of high artistic value. The Forty-Eighters thus awakened interests which a majority of the old population had hardly known, but which now attracted general favor and very largely bridged over the distance between the native American and the new-comer.

The establishment of a German theater was a matter of course, and its performances, which indeed deserved much praise, proved so attractive that it became a sort of social center.

It is true, similar things were done in other cities where the Forty-Eighters had congregated. But so far as I know, nowhere did their influence so quickly impress itself upon the whole social atmosphere as in the "German Athens of America," as Milwaukee was called at the time.

Ernst L. Meyer recalled the feeling of Gemütlichkeit *(geniality, which he described as "serenity of the mind and a physical sense of expansive well-being") created by the many different German American* Vereine. *He described those in Milwaukee, where he grew up in the early 20th century.*

With the coming of fall the good people of Milwaukee slid into a routine that was well-worn, comfortable, and above all communal. They functioned in groups, and when one is part of a group one does not do the extravagant things that one does standing apart and alone. In the fall, the dormant life of the thousand clubs and societies reawakened; there were concerts and theatre parties and skat [a card game] tournaments and gymnastic competitions—a hundred things that clamored for attention....

[On Wednesday nights there were special performances at the Pabst Theatre.] We buy season tickets each year, and from October to May go each Wednesday to sink in red plush and weep and laugh before the footlights....

On Saturday and Sunday nights they put on frothy bits—

light operas, comedies, farces—but on Wednesdays they don their buskins and tread the stage in graver measures. Goethe, Schiller, Shakespeare, Strindberg, Ibsen, Von Hoffmannsthal, Hebbel, Schnitzler, Shaw—what a roster of great names and great plays, what nights to be remembered.

Tonight everyone feels especially proud. We are going to see the first of the *König Heinrich* [King Henry] trilogy. It is ponderous, it is slow moving, but it is heroic and soul-filling—a dramatic joint of beef.... It has fifteen acts and will take three nights to play. Three nights!...

So we settle in our seats in the balcony amid a buzz of chatter. Everyone knows everyone else, for this is subscribers' night. Each Wednesday night we know exactly who will sit to our right, to our left, and behind and in front of us. Yet each Wednesday we shake hands solemnly with our neighbors and launch a cannonade of small talk as if we had not seen them in seven years. My mother says, "Ach, Frau Braunschweig, it is so good to see you again. I really must come and call on you one day."

"On my next *Kaffeeklatsch* [coffee party with friends], Frau Meyer, I shall most certainly call you on the telephone. You have never tried my Zwiebelkuchen, which is my specialty. So this is your little daughter Freda? What a pretty child!..."

So we chatter before the curtain goes up, and there is a running about as people flit up and down the aisles to shake hands with old cronies, former sweethearts, business friends. Old Hermann Suelze is the only one who doesn't budge. Hermann weighs perhaps three hundred pounds.... He subscribes each season to two seats, all for himself, to take care of the overflow.

There is a certain constraint or self-consciousness in the lower boxes. Here, on gilt chairs, sit the German-American

As Others Saw Them

In 1855, a reporter for the New York Times *described "Dutch Hill," a district of New York City along the East River which is now the site of the United Nations:*

Some [of the shacks] are of the primitive block form, with a hole in the roof for a chimney; others are arched, others with a sharp Gothic gable. Occasionally something entirely new in architectural style will meet you in the shape of a rectangular box with diamond lattice work, which, on nearer approach, you discover to be a railroad car...made into a house.... Each house has a retinue of goats and pigs.... All the inhabitants of these buildings are squatters—they have found a plot of ground and have built their log cabin on it, to remain until the rightful owner turns them away. When they move they sell their house to some newcomer for $5 or $10. They are all Irish and German laborers; many of them working in the quarries nearby, and others, especially the German women, living on the sale of the rags and bones which they and their children gather all the day long through the streets of the City.

Picnic *is a German word (spelled* picknick*) that has entered the English language. German Americans expressed their love of nature by going outdoors to enjoy* wursts *(sausages) of all kinds and the ever-present glass of beer.*

The Turnhalle, *or clubhouse, of the* Turnverein *of Lawrence, Massachusetts, around 1916. It included an elaborate stage and dance hall where members could enjoy food and drink while watching German-language plays and musical performances.*

A Maennerchor, *or men's singing society, named after one of the great German-speaking composers.*

elite, the great ones, who are patrons of the theatre and who yearly contribute big sums so that Milwaukee shall have a stage of which one can boast. That fine dame in satin, for instance—well, she inherited her wealth from Captain Joseph Schlitz who came to Milwaukee no more than fifty years ago and laid the foundation of his fortune by brewing each day three barrels of beer and peddling them himself in a wheelbarrow. Too bad the poor barrow-pusher is not here tonight to see the grand theatre which his money helps support, and to feast his eyes on the lady in white satin. But no man lives forever, not even though his name, which "Made Milwaukee Famous," still blazes from posters that shadow his tomb....

Then the curtain rises. Everyone is glad that Konrad Bolten and Berthold Sprotte are cast in major roles. They are prime favorites.... They make friends everywhere. Why, they have often visited at our own house, where they sang and joked while I stood goggle-eyed in mute admiration of the deep-chested Berthold who only a week before I had seen slain in *Macbeth*. And there he was, lively as you please, stuffing great mouthfuls of food and complimenting my mother upon the excellence of her *Herringsalat*.

In the long intermission, all the men arise and file downstairs to the Pabst Buffet, while the women and children remain behind and fill the theatre with shrill voices. In the buffet there is Pabst on tap, and excellent free lunch. Two minutes before the curtain rises a bell clangs loudly in the buffet, and the men file upstairs again, wiping their whiskers.

But after the performance it is the women's turn. The men go with them, some of them reluctantly, to Martini's a half block from the theatre. It is a confectioner's shop specializing in German dainties: *Schaumtorte, Apfelkuchen, Kartoffeltorte, Marzipan,* thick chocolate with whipped cream, and strong coffee.... After a bit, some of the actors come, and the groups at the tables fight for the honor of claiming them. The long, narrow room buzzes with talk; laughter cuts through the fragrant haze of cigar smoke and the steam arising from the cups.

Oscar Ameringer lost his job in Cincinnati during the winter of 1887. In his autobiography, he described how the beer halls' free lunch for customers kept him alive.

The problem of eating was solved by the greatest and most beneficent of all American institutions—the free-lunch counter. By investing five cents in a schooner of beer and holding on to the evidence of purchase, one could eat one's fill of such delicacies as rye bread, cheese, hams, sausage, pickled and smoked herring, sardines, onions, radishes, and pumpernickel. It is true I did not always have the required nickel. But by patronizing only the larger saloons through the rush hours one could always commandeer a partially filled glass some absent-minded cash customer had left unguarded, and by doing so escape the suspicion of being a deadbeat.

Immigrant Richard Bartholdt described the first "German Day" in St. Louis. While editing a newspaper there he helped to organize this celebration of German Americans' heritage, to counter anti-German prejudice in the 1880s.

Street parades were to demonstrate the numerical strength of the German element and orators should call public attention to its creditable history on American soil. And the special event to be commemorated? Was it to be chosen from the history of the old country? No; these citizens of German blood wished...to demonstrate their Americanism. Hence, they decided to commemorate the landing here 200 years earlier, of the first German colonists—an event of importance in the history of America.

The idea spread like wildfire over the whole country, and for many years thereafter annual festivities of this character took place in all large cities and also in many smaller towns.....

German American women formed their own Turnvereine *because physical fitness was regarded as a goal for both sexes. The wooden clubs were used in a variety of exercises to develop both strength and dexterity.*

These apartments are certainly small, but are therefore more [efficient]. A parlor with two windows and...a windowless bedroom, in which there is room for a double bed—and nothing else! Closet, kitchen, cellar, woodshed or the like is unknown; not even a corridor or pantry is to be found. Overall—how large is the parlor with the two windows? Ten feet long and ten feet wide, seldom more!... [This leaves no room for children and if one needs more room] it is necessary to rent in a house with ten or twelve families.

At the first celebration of this kind at St. Louis there were more than 40,000 men in line, each carrying an American flag. As one of the collaborators I was not a little proud of the success of the turnout.

Lucille Kohler remembered her childhood in St. Louis in the early years of the 20th century. She described the short season soon after Easter when breweries produced bock beer, a stronger, darker beverage than lager beer.

We knew that while bock beer lasted the *Eltern* [elders] would be gayer, kinder. We knew that while bock beer lasted pretzels would be free at all beer saloon counters, and patrons, moved to song, would grow hoarse in *Sangerfests*. We knew that while bock beer lasted, there would be many who would marry, some even for a second time; and second weddings were twice as much fun. We knew that with bock beer and pinochle the grown-ups would let the evenings stretch and give us our fill of games and peanuts....

From after supper until dark we might follow a Little German Band from beer saloon to beer saloon in our neighborhood, listen to the singing, and reap pretzels and soda water...we attended charivaris [shivarees, or mock serenades for a wedding couple], pinochle and *klatsch fests* [gatherings for a festival], a concert at Liederkranz Hall, and never did we see our beds before nine, even ten, o'clock.

The Germans reveled in their public festivals. One of the most popular was the Cannstatter Volksfest *or fall festival, celebrated in Chicago by the* Schwaben Verein. *The* Chicagoer Arbeiter-Zeitung *described a celebration in August 1879.*

The German marketplace in Milwaukee around 1896. Large numbers of German immigrants have settled in the city ever since its founding in 1836. Most lived on the east side, where virtually all business and trade was carried on in German.

Yesterday's questionable weather, which was quite threatening for a while, couldn't prevent hordes of Chicago's merry Germans from going to Ogden's Grove on Clybourn Ave. for the annual Swabian fall festival. A cloud of dust hovered over the fest site itself, and even from far away it was evident that a huge crowd had gathered there. This was confirmed again upon entering the "garden," or grove, where one caught sight of the thousands of visitors being swept along by the maelstrom, helplessly tottering back and forth amidst the powerful human current. The conglomeration of folk culture which was developing here was, as always, of an extremely interesting and enjoyable nature. The carefree observer was offered a tableau of such earthy and exciting activity that it was a pure pleasure simply to be in the middle of the flow. The usual platitudes about the success of the festival cannot be applied to this occasion, as it poses an exception to the rule. And although this reporter is used to a lot, he nevertheless always finds new material here for original comparisons.

The arrangement of the stands, booths, tents for eating and drinking, the carousels, theater, and other attractions are generally the same as they have always been, but this is the very least. The main attractions are the stage presentations, the performances by amateurs and artists, the outstanding promenade music, folk dances, illuminations and fireworks, shadowgraphs, etc., and finally the faithful rush to the various booths, the most popular of which is the wine bar, the social center for the cheerful Swabians. This [is] also where the magnificent "pillar of the vine," under which the 1873 Riesling from the royal caves in Stuttgart is sold, has been set up.

Yesterday's Cannstatter Volkfest was wonderful, and as today is yet another day promising good times under the "pillar of the vine," it will presumably be no less so. All of yesterday's performances will be repeated today, so whoever wants to participate in the last day's activities, just come along to Ogden's Grove. You won't regret it.

The worst tragedy to hit the German community in New York was a fire on the ferryboat General Slocum *on June 15, 1904. More than 1,000 people died on an excursion planned for members of the St. Mark's German Lutheran Church. Edna Doering, a survivor, recalled the tragedy many years later.*

The German community was more around 8th Street. That's where St. Mark's Lutheran Church was. My father was a Lutheran minister. He was a director of an immigrant house in that neighborhood.

In 1904, my brother was nine and my sister eleven. I was only six. That day, we were going on a Sunday-school picnic on Brother's Island. That's where we were headed for. Of course, we never got there.

I remember the excitement going to the ship. We got seats with my mother right near the railing. She brought a nice big

As Others Saw Them

In summer evenings, the residents of New York City's Little Germany sought relief from their stuffy apartments by gathering in the streets. Musicians performed for coins, and vendors sold beer and other refreshments. A German American writer described one such scene in the 1880s:

Even the children devoted their complete attention to the band of musicians in the middle of the block.... The young people, despite the heat of the evening, roused themselves to an impromptu dance everytime the band struck up a waltz. The boys and girls would dance through the openings in the ranks of spectators. It was the kind of night the musicians loved—when people shuddered at the thought of bed and, as long as their tired eyes held out, camped out in the open air of the streets where, along with the pleasures of music and dance, the beer jug made the rounds. On such nights the beer sellers and street musicians were in clover.

A German American grocery store around 1900. As the largest ethnic group in the United States, German Americans could support stores that sold their favorite foods.

This was not a club for those who liked to eat ground-beef patties but an association of people who came from Hamburg. Many similar groups were formed among German Americans who came from other regions of Germany.

shiny tin box for our lunch. In the initial excitement, when they called "Fire," she threw it overboard and I felt so bad that our lunch was being thrown over.

My mother tried to get life preservers but they were rotten. I remember some of the cork actually falling down over her. Then my mother had my hand and she tried to go to the upper deck. But it was so hot that she turned around. At that point we got separated. When the people heard "Fire!" they rushed toward the railing and broke it off. The next thing I knew I was standing at the opening. There was a woman at the other end sitting, and she came over and threw me overboard. I don't know who she was, and I never saw her again, but I owe my life to her.

I don't remember ever being afraid, and I don't remember falling. I remember the feeling of being in the water, and there were so many bodies I just kept pushing. I guess that's what kept me alive. I was floating on the bodies.

Then, my mother, who had already been taken in a rowboat, spotted me, and she asked the man please to get me, so I got in the same rowboat with my mother. The next thing I remember I was on a barge, and a man had me in his arms and put a blanket around me.

At first I was put with the dead, and a doctor came along and he saw there was life. I was put in an ambulance and taken to a hospital. There was a man there who worked in the immigrant house. He came to the hospital where I was and he saw me. He asked me where my mother was. I said I didn't know. A short time later he came back with her. She was in the same hospital.

My mother lived six days after the accident. She was badly burned and she had pneumonia. My brother was found the next day. Of course he drowned. My sister wasn't found until six days later. By that time, she was decomposed and it was hard to identify her. An uncle of mine went out every day looking for her. They laid the dead in rows. Another woman was claiming the body at the same time, but my uncle came home [and] asked where different articles of clothing were bought, because they found some markers. The big store in those days was Siegel-Cooper. My mother said the shoes were bought there, and that's the only way they identified my sister.

My mother was very sick. My father didn't want her to know that the other two children were dead. That's all she did for six days—call to her children. I remember one day there was a lot [of] commotion, and somebody told me my brother was being buried that day, the same day they found my sister. They didn't even bring her to the house. She was buried from another immigrant house....

You could imagine the poor little church. It had over a thousand people in the Sunday school, and after that it dwindled down to almost nothing. The poor minister—I can't imagine, his whole congregation. They lost 1,500 people on the *Titanic*. We were 1,030, and when you think of 1,030 people—especially from one neighborhood—it's hard to picture that.

Singing societies were part of virtually every German American community. The societies enjoyed meeting with other groups in festivals called Sängerfeste. *Emma Murck Altgelt described the Texas state* Sängerfest *held in New Braunfels in 1855.*

Many singers came from far away, among them a jolly group from Sisterdale.... They came by oxteam because the day before, Indians had stolen all of the horses in the settlement. The suits of most of the men were quaint. Although they were aristocrats, not every one of them had a coat. The German apparel was worn out, little of it was new and that which was bought was inferior. A large singers' hall had been erected. The ladies furnished the flowers and wreaths. Everyone brought good humor, and many beautiful songs were beautifully sung.

Elsie Mohr was born in Nebraska in 1891 and her family moved to Minnesota when she was eleven. She told an interviewer in 1974 about Horton Township, Minnesota.

Ya, ya, everything was German. Wherever we went, everybody talked German. I never knew any different. There was very few, a few Irish people and that was about all that was around, that was all German, ya....

German Americans celebrated Maifest on the first day of May. Here, women with parasols in a flower-bedecked carriage participate in a Maifest parade in New Braunfels, Texas.

Members of New York City's Liederkranz Society are dressed in costumes for a play around 1898. Stories and legends from German history were popular subjects for drama.

Students at the Indiana State Teachers College demonstrate their gymnastics skills. The Turners, founded in Berlin in 1811, developed such gym equipment as the parallel bars and vaulting horse.

Members of the Los Angeles Turnverein gather in front of their newly built headquarters in 1888.

There was German clerks in all the stores, you know. Don't make no difference all the business places where you walked in they could talk German.... When I was a young girl, we never had to talk English. It was all German, regardless of where you went in the business places, they all had German clerks that was no bad thing....

People it seems like, I know, didn't have much, but everybody was so happy. You know I can remember the folks and my folks they'd go together on Sunday afternoons, they'd have a party and they'd have a keg of beer and those men would sit around under the trees in the summer time, and they had their beer and they'd sing songs or they'd play cards. And the women too, they really enjoyed it. Everybody was so neighborly you know. We never knew of not getting together.

Lawrence Welk grew up in the German American community of Strasburg, North Dakota, in the early 20th century. He recalled the feelings of pride that the annual Fourth of July parade there engendered.

When I was a small child, our whole family had always gone into Strasburg for the day, dressed in the best clothes we had, laden with a huge picnic lunch, ready to celebrate. My father gave each of us smaller children a dime apiece to spend—more than we ever got at any other time of the year—and I was always in [an] agony of delight trying to decide just how to spend all that money. Mike and I would spend hours trying to make a choice between buying candy or lemonade or fireworks or maybe saving a penny or two....

Everybody gathered in town during the early part of the day, and in the afternoon the parade would start forming at one end of the main street. The little Strasburg band would come down the street playing a march for all it was worth, and everybody crowded to the edge of the sidewalk to watch it go by. The band always marched at the head of the parade and the town fire engine brought up the rear, with all the local dignitaries riding in floats or cars in between. And always, at the head of the parade, was the American flag, fluttering in the breeze, making small crackling noises as it passed by the rows of silent people—the men with their hats over their hearts, the women standing very straight and quiet, the children saluting. It was a hushed, almost reverent moment, a time of great emotion for all of us. I have often thought that those immigrant farmers, most of whom could not even speak English, were probably the most American Americans I have ever known. They appreciated the wonders of this country, more perhaps than those who had been born into it. They had all experienced deprivation and tyranny in the old country and they never got over the freedom and opportunity they found here. Deeply grateful for it, they taught all of us children to value it as they did.

Kleindeutschland, called Dutchtown by the Irish, consisted of 400 blocks formed by some six avenues and nearly forty streets. Tompkins Square formed pretty much the center. Avenue B, occasionally called the German Broadway, was the commercial artery. Each basement was a workshop, every first floor was a store, and the partially roofed sidewalks were markets for goods of all sorts. Avenue A was the street for beer halls, oyster saloons and groceries. The Bowery was the western border (anything further west was totally foreign), but it was also the amusement and loafing district. There all the artistic treats, from classical drama to puppet comedies, were [available].

—German American writer Otto Lohr describing New York's "Little Germany" as it existed in the 1870s.

Cincinnati's "Over-the-Rhine" neighborhood around 1890. Like Milwaukee, Cincinnati was a center of German American life. David Ziegler, the city's first mayor, was a German. In 1890, more than half of the city's residents were first- or second-generation German Americans.

FAMILY

Jacob Gross arrived in Chicago in 1855 with his wife and children. Everyone in the family did their part to help make a living. In the following letter to his brothers and sisters in Kadelburg, Gross describes the activities of each of his seven children, along with his own and his wife's.

As to our employment: I and Theodor make wood in the city…. Last summer we made hay on the prairies—a very good business to make money. Theodor wanted to learn a trade, but I could not spare him. Leopold is learning plastering and whitewashing and gets 10 shillings or 3 gulden a day, but without board. Otto is learning the butcher trade and gets 12 dollars a month and board—he is strong and has grown large. Jacob goes to school and on the side helps to slaughter [in a butcher shop] and earns 2 to 3 shillings a day and enough meat for our own use. Elise also visits the school when it is not too cold. Elizabeth was [working] in a hotel until this fall and earned 12 shillings a week, but this winter she is at home learning to sew and iron so that she will be ready to take an English position in the spring. Marie and the wife do washing in the town—mostly for Kadelburgers. We all work and have good incomes, but also plenty of expenses.

Immigrants soon found that relationships between men and women were different in the United States. In Germany it was customary for a wife's family to provide her with a dowry, money to help her and her husband get started. In the New World, however, customs were different, as seen in this 1853 letter from Carl Hilmar Guenther, who had emigrated to Texas. He wrote to his father in Germany, describing his prospects for marriage.

Dear Father, you suggest that in choosing a wife I should keep an eye out for one with a good dowry. But in this country, I would be glad if I could find one that I could like and forget the dowry. Life with a housekeeper here would be distasteful to me, because I could not find a woman who would be suitable. Every nice girl gets married here at an early age, so that leaves only the undesirable ones and one of that kind could not be an efficient housekeeper. There are enough and pretty girls here, but the cultured and educated ones are rare and it would be quite impossible for me to marry an uneducated person. Here in America a man's wife can influence…his career much more easily than in Germany, because here everybody depends entirely on himself. Even if there were a girl here with a dowry, she would be so conceited about it that it would only cause trouble.

A card game in the Dick family's backyard grape arbor in Milwaukee around 1900. Though Protestants of English descent frowned on such activities, German Americans regarded card playing and beer drinking as harmless enjoyments.

Ernst L. Meyer lovingly recalled his immigrant mother's determination to keep their house clean and healthy.

Cleanliness, I fear, was a bit of a fixation with her....
Every morning precisely at ten o'clock, for example,
mother would throw open all the windows in the
house, no matter what the season, for a thorough airing....

Her daily routine was matched by rituals in the procession
of seasons. At an almost unvarying date in early April, each
year, mother ordered the coal stove [taken apart for cleaning.
It had] a stovepipe hung on wires along the ceiling for almost
the entire length of the house. This was a time of anguish for
father and me, because neither of us was good at things mechanical, and as a rule a length of pipe would come clattering
down and the soot would get into father's beard and mother's
favorite rug, and father would explode into a roar of profan-
ity. He was good at this, for he had received a sound education
and could swear in German, English, Latin, and Greek. When
we children were about, father quite properly on these occasions swore only in Latin and Greek, lest our ears burn....

Every fall, with a regularity only the frugal know, there
was a long season of pickling and preserving, of laying down a
whole barrel of Sauerkraut, of filling hundreds of quart jars
and glasses with jams and jellies and compotes and cucumbers
and relishes and catsup, meant to last for a whole year lest
mother suffer the humiliation of having to serve, at supper or
at her weekly *Kaffeeklatsch* [coffee party] preserves shamefully
dug out of tins purchased at the corner grocery and which every good *Hausfrau* [housewife] of that era could detect at
once, sniffishly.... My sisters and I...had to assist mother
bravely in the picklings. It was my especial task, as being the
strongest, to squeeze out to the last drop the juice of the
crushed grapes mother put into a large square cheese-cloth tied
to the legs of a stool which had been placed upside down on
the kitchen table. Though I liked the heady, winy smell of the
dripping fruit, I did not relish the job much, for it left my
hands stained for days.

The bride in this wedding portrait from Michigan wears a black dress, which was the custom among many of the German Protestant groups.

A German American picnic in Boonville, Missouri. Farther down the Missouri River, St. Louis was the third major midwestern city where German American culture was dominant.

A birthday party at the Limberger house in Milwaukee around 1910. The cake has little American flags on it, signifying the patriotism that was a strong feature of German American life. Less than a decade later, however, during World War I, the Limbergers changed their name to Lembert, possibly as a reaction to the anti-German prejudice of the time.

Sophia Heller was less than a year old when her parents left Bohemia in the late 1840s. They settled in the newly established community of Milwaukee. When Sophia was 13, she first saw Phillip Goldsmith, the young man destined to become her husband. Four years later, Phillip obtained Sophia's parents' permission to marry her. The young couple attended a dinner of celebration with their friends and relatives. Sophia recalled this event in a memoir she wrote for her children.

We were all seated. I was quite embarrassed, like a schoolgirl, so shy. Father asked Mr. G. if he would have me for his future wife before everybody; then asked me if I would have Mr. G. I turned all colors, not expecting this. After answering, my father said: "Kiss each other." This was quite embarrassing. Mr. P. G. placed a ring, rather three rings in one, "two hands over two hearts and same was closed," it was the emblem of love on my finger, and he kissed me.

At a very late hour we broke the gathering up. Mr. Stumas, the intimate friend of P. G., took us home at two-thirty in the morning from this celebration. On the way he took two hands full of silver coins and threw it in the air for the luck of the young couple.

This happened the second [day] of Chevuos [Shavuos, the Jewish festival that commemorates the giving of the Ten Commandments to Moses], the latter part of May, 1865.

Sally Roesch fondly recalled her grandmother, Rosina Treftz Roesch, who emigrated to North Dakota. As a girl in Russia in the 1870s, Rosina Treftz had a teacher who teased her by making a pun on her last name, which means "catch it" in German. Her granddaughter explained the wordplay.

Treftz nicht immer" ["You don't always catch it"], her teacher told her. I would be sitting at her feet in the upstairs bedroom, where she stayed with us part of the time during my childhood. Back and forth in her rocking chair, combing her long, black-streaked gray hair, she would laugh anew as she repeated the story for the umpteenth time. My eyes were riveted on her surprisingly agile fingers, holding the tortoise shell rounded comb like a small mouse in her ham of a hand, deftly drawing teeth so fine and thinly-spaced that they seemed impenetrable, through her thin hair. Then she would quickly weave a short strand of black material into a braid on either side with two bone hair pins. Her never idle hands picked up her knitting.

Laughing over earlier times as her fingers flicked back and forth, the crochet hooks my Uncle Ed had whittled for her circling the growing outline of the small rug she made, every day a new one, out of four skeins of yarn. Everybody got a rug when they came to visit her, and still there were cabinets full of them when she died. One of mine finally fell to pieces from wear and washing; the other two warm my feet in front of the reading chair in my study.

Helen Wagner grew up in Yorkville, a German American neighborhood in New York City, in the early 20th century. She recalled dinnertime there with her family.

W e always ate German food, a lot of beef, sauerbeef and meat loaf. At night we ate a full course supper, and rest assured you better eat it. It was soup and potatoes and vegetables and meat and some kind of a cooked dessert.

We were permitted to talk to one another until it was time to eat. Then we had to keep quiet and eat. My dad always said you couldn't do two things right at one time. If we wanted to ask my father something, that was permitted, but we couldn't converse with one another at the table....

Sundays was my father's treat to take me to Central Park with a bag of peanuts and something for the pigeons, and we'd walk all the way down to 59th Street and all the way back home again. We would stop across the street, and I would be treated to two Cel-Ray tonics and my dad would have two glasses of beer. My brother would meet him and we'd go up with the growler [a bucket of beer], and that was our Sunday relaxation.

German American families proudly preserved the German language and customs. Hermann Hagedorn, born in 1882 to German immigrant parents in Brooklyn, New York, recalled those times.

I t was a German world in which we lived. Whether it were in the pleasant house of Berkeley Place, or the narrower, dingier house on Hancock Street, in the Bedford Heights section, it was as German as Bingen on the Rhine. The girls and I generally talked English with Mother, but, when Father was around, and especially at the dinner-table, the language was German. Occasionally, if we were forgetful or obdurate, slipping into English, or returning to German too tardily, the paternal fist would make the dishes rattle with an emphatic but good-natured *"Hier wird deutsch gesprochen!"* ["Here, German

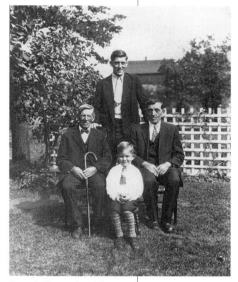

Joseph Beranek with his son, grandson, and great-grandson in LaCrosse, Wisconsin, around 1920. Proud of their traditions, German Americans passed them on to later generations.

A Mennonite family in Lawrence, Kansas, in the late 19th century. The first German American immigrants were Mennonites, and others continued to arrive throughout the 19th century. Today, there are more Mennonites in the United States (about 250,000) than in any other country.

The Stratemeyer Family

Few, if any, American authors have been as popular as Harriet Stratemeyer Adams and her father, Edward Stratemeyer. They are better known, however, as Carolyn Keene, Franklin W. Dixon, Laura Lee Hope, Victor Appleton, or one of the many other pen names under which they published their books.

Edward Stratemeyer was born in 1862, the son of a German immigrant who had arrived in California during the gold rush of 1849. Although Edward was an avid reader, he could not afford to go to school beyond the eighth grade. One day, while working in his brother's tobacco store, he began to write a story on a sheet of brown wrapping paper. A magazine bought it for $75, six times his weekly pay in the store.

That was enough to convince him to write full-time. He started with "dime novels," inexpensive paperbound books that appealed to young readers. Though teachers and librarians frowned on the slam-bang action, Stratemeyer had millions of readers. He hit on the formula of creating series with young heroes and heroines, such as the Rover Boys, Tom Swift, Ruth Fielding, and the Bobbsey Twins.

Stratemeyer could write 7,500 words a day, but publishers demanded more. He began to outline books and hire others to write them. In this way, he produced more than 1,300 books.

His daughter Harriet, a graduate of Wellesley College, took over the family business when her father died in 1930. She concentrated on her two favorite series, Nancy Drew and the Hardy Boys, which her father had begun just before his death.

Harriet Stratemeyer insisted that each book should teach a moral and have some educational content. The characters Nancy, Frank, and Joe always learn intriguing facts on their way to solving the mystery. By the time of her death in 1982, young readers had bought more than 250 million copies of the Nancy Drew and Hardy Boys books.

is spoken!"] Talking German seemed to us somehow strained, but Father was right, of course, when he said that someday we would be glad he had been strict about it.

Lawrence Welk, who would one day entertain audiences of millions with his accordion, started his career by playing at weddings in his hometown of Strasburg, North Dakota. In his autobiography he remembered the festivities of a typical wedding.

Since the community was almost entirely Catholic, the wedding mass was usually held at St. Peter and Paul's Church at ten o'clock in the morning, and then the wedding party and all the guests would return to the bride's home for the festivities. My part was to stand near the door and "play" the bride and groom into the house.... After the wedding party was safely inside, I rushed around to the wedding table and began playing dinner music....

Those wedding feasts certainly offered food to be remembered! Platters of fried chicken, roast beef, ham, sauerbraten, wurst (a German sausage), all kinds of home-canned fruits and vegetables and big bowls of sauerkraut, home-baked bread and rolls and German chocolate cakes and butter cookies and always a special confection called *"Hochzeitkuchen"* or *"Kuka,"* a flat cookie-like cake which was heavily sprinkled with sugar and cinnamon and served traditionally in pie-shaped wedges at all German weddings.... It always seemed a wonder to me that the guests could even get up and walk, let alone dance, after such a feast. But dance they did, and that was the part I loved best.... In most cases [the dances] were held in the hayloft of the barn, which...could handle as many as a hundred couples or more. The haylofts were always sturdily built...so they could support the heavy weight of the hay which was stored there during the winter. And a good thing it was too, or they could never have supported the weight of all those dancers. These people didn't glide sedately around in a smooth waltz or fox trot. They *danced*, with a good deal of heel stomping and foot clicking. After an hour or so of these whirling shouting dances, the atmosphere got so thick it made your eyes water....

Whole families came along to these parties, and children were sometimes lost among the piles of coats and sweaters on the benches, sound asleep as their parents spun and danced around them....

It was not unusual for the festivities to go on well after midnight the first night. The next day the same thing started all over again. By the end of the second day many of the older folks bid the bride and groom goodbye and set off for home. But the rest of the guests, sometimes with a quick trip home to attend to necessary farm chores, kept on going.

The relatives of Andrew Wieber gather for his wake in Stearns County, Minnesota, in 1912. Though his body was wasted by tuberculosis, Wieber was only 27 when he died. It was common for families to have a final picture taken of the deceased.

In 1978, Cecelia Kost Butala of Richmond, Minnesota, told an interviewer about growing up on a Minnesota farm in the early 20th century.

We had our big garden. We raised all our own vegetables. [Mother] did all the canning, about 800 quarts. We made our own sauerkraut in the big barrel and dill pickles in a big barrel. Carrots were put in the sand and that way we had all our own vegetables for the whole year. Plenty potatoes. Popped lots of popcorn. That's what we used to serve when we had parties. Popcorn and lemonade. Dancing in our big kitchen with an old talking machine going.

It was lots of fun. All the young kids, instead of going out and having parties, we had it in the house. We had parties about every Saturday night I remember in our big kitchen....

Q: Can you tell me about sausage making?

Well, years ago, the farmers would all get together...and we'd butcher about four or five hogs at one time. Sausage was made the same day. And pigs were cut up, put in brine and afterwards smoked. We made blood sausage, liver sausage, and other pork sausage.... We kids just loved it when people would get together...until way in the night and when the sausage was done we had a sausage fry.

Archie and Clarence Bisenius in front of the Christmas tree at their grandparents' home in St. Cloud, Minnesota, in 1906. The custom of decorating a tree for Christmas came to the United States with German Americans.

A quilting bee at the Methodist Episcopal Church Society in Lawrence, Kansas, around 1890. The quilt might have been sold or raffled to raise money for the church.

Placing an elaborate iron cross to mark a grave site was a tradition brought to the United States by Germans from the Volga River region of Russia. This one is at St. Fidelis Cemetery in Hays, Kansas.

RELIGION

C. C. Wedel, who came to Kansas in 1874 as a young boy with his family, described the church services of the Mennonite community there.

The benches were of plain boards nailed to wood horses without back rests. A narrow aisle was left in the center and benches around the walls which were the most comfortable ones were always taken first. The services usually lasted from two to two and a half hours. The four or five songleaders announced two or three songs with a great number of verses which were sung without instrument accompaniment. Then came the sermons, first one preacher made the opening which was longer than the sermons we have today followed by the regular sermon; so it was always a long service.... They often had a recess between the opening and sermon which was taken advantage of by many people to go out for a little while....

I was a barefoot boy and as I saw people leave the church, I thought the service was over with. I went to our wagon and waited a while but nobody came. I did see that one by one the people went back into the service but I didn't think too much about it. When we went home from church, I got straightened out as to what was the right thing to do

Hermann Hagedorn, born in Brooklyn in 1882, recalled the importance to his family of the Lutheran Schermerhorn Street Church.

The church loomed large in our lives. Father was fond of its minister, the genial, bewhiskered Reverend Jacob Loch. "Jakey," as we children irreverently dubbed him, loved a good glass of wine and an evening of cards, but his heart was sound and he was a hard-working, kindly soul; and Sundays he was all gravity and unction.

Pastor Loch confirmed Elsie, Irma and myself, in succession. Confirmation in those days was a serious business, preceded by three or four months of weekly lessons in Christian doctrine and Lutheran theology, culminating in a public examination before the whole congregation. This ordeal was actually not as grim as it sounds, for the test was rigged by the parson to prevent any failure, embarrassing alike to the confirmee and the congregation. "Jakey" knew what each of us knew and didn't know, and adjusted his questions accordingly. The gravity of the sacrament of confirmation was impressed on me on Palm Sunday morning by Father's bringing the big German Bible to breakfast and reading his favorite psalm, the 103rd.

Sophia Goldsmith's family settled in Milwaukee around 1848. Ten years later the Jewish community of Milwaukee had outgrown its original synagogue and raised funds to build a new one. Many years later, Goldsmith recalled that Dr. Isaac M. Wise, the leading figure in American Reform Judaism, dedicated the synagogue.

These women enjoy a church picnic in Milwaukee around 1895.

I was then about ten years of age.... The Jewish community was in one excitement to have such a learned man come from Cincinnati to our city. He was invited to our house and his stay was a short one. He only called, and how well I remember him as a young man. Two girls and myself were the happy children robed in white with baskets of flowers. We walked in front of Dr. Wise as he carried the *Safra-torah* [Torah scroll] around in the temple. We strewed flowers in front of him.

Mother had presented the temple with a *Shulchendeck* [altar cloth]. Same was made of red silk velvet and embroidered in gold. Father had bought the best seat in the temple. How big I felt! It was quite an honor to see my parents so prosperous. About seven years later I was married in the same temple.

Angela Heck emigrated from Prussia with her husband Nikolaus in 1854. They settled in Williamsburg, today part of Brooklyn, New York. Like most of their neighbors, the Hecks were Catholics. Angela Heck wrote to her relatives in Germany in 1869, describing the role of the church in their lives.

All of the small shops are German, 17–18 German-Catholic churches and each one has a Catholic school. We don't have to go further to church than you do at home.... There's a big school there, with 8–9 teachers, and also that many nuns. The nuns teach the girls and also the little boys until they are grown up. Then they go to the high school until the age of 14. Our little Nikolaus also goes there. We pay 15 cents a week. In the [high] school there are 1500 [students], all with German parents and all Catholic; the priests visit the schools every day and are very strict. We also belong to this parish. We have a pew in the church. For that we pay one taler every three months. My husband belongs to the St. Joseph's society. If he gets sick, he gets 4 talers a week sick pay. And if he dies, he will be collected from the house and taken to the churchyard by the members of the society. The society covers all the expenses.... We have three priests in our parish. One arrived here a short while ago. One is a Bavarian, one is an Alsatian. There is confession every Saturday. They are much stricter with confession and sermons here than in Germany.... In all of the churches there are lovely organs and four-part choirs, all German singers.

Gilbert J. Jordan remembered the Sunday school he attended in the Methodist church in Plehweville, Texas, around 1910.

The sessions opened with a general program of singing, reading, prayer, and an exhortation by the Sunday school superintendent. The songs were mostly lively

The congregation of St. Peter's Lutheran Church, near Hazen, North Dakota, in 1914. Typically, members of German Russian congregations were segregated by sex in the church services. Women and girls are at left, men and boys at right.

The confirmation class of 1900 at Temple Beth El, a Jewish congregation in Providence, Rhode Island. Jewish confirmation ceremonies (replacing the more orthodox bar mitzvah ceremony) originated with the Reform branch of Judaism, which was brought to the United States by Jews from Germany.

and appealing to young people. The little children were taught to sing several songs in German. Our favorite was "Gott ist die Liebe; Lässt mich erlösen" ("God is All Loving; He Would Redeem Me")....

On certain Sundays of the month...the young people had evening Epworth League [Methodist youth organization] services and occasionally a song practice.... Dating also played a role in these gatherings. The boys took this opportunity to come by for their girl friends and take them back home in their courting buggies. From 1915 on, they used the family cars.

While growing up in North Dakota in the 1940s, Emma M. Boardman belonged to the Mennonite Brethren Church. In 1985, she described some of their customs and practices.

One was born into the church, of course—few outsiders sought membership—but there was also a powerful belief that no individual was truly a part of it without a specific conversion experience.... This was the born-again experience, often called "getting saved," and a vivid, unforgettable milestone for each of us. After it, we were addressed as Brother—or Sister—rarely by name alone within the church. We were not a rebellious generation, and most of us embraced the experience in our early teens....

Baptism ("believers' Baptism") was an experience no one forgot. A warm July Sunday afternoon and a small lake with a sandy bottom were the time and the place.... We sang unaccompanied from hymnals held open against the wind: "Shall We Gather at the River" and "Wash Me and I Shall Be Whiter Than Snow." The minister...spoke of [Baptism's] cleansing symbolism, signifying the death of the natural, sinful man and the rising to life of the reborn man. Then he would wade into the water waist deep, ruining a pair of pants in the pre-polyester age. One by one, we would follow. Our age range was from about 12 to adults of all ages. Few of us had had any experience with swimming, so the experience of wading into deep water was unique and a little frightening. Our clothing would be common, cotton garments, and each of us would carry a large, folded men's handkerchief to keep the water from entering our noses. The water, at this time of year, was smelly and a light green from algae. Clutching us firmly, one hand supporting our back, he would tip us once backward into the sun-warmed water. "I baptize you in the name of the Father, and of the Son, and of the Holy Spirit. Amen." Dripping and disoriented, but exuberant, we waded back to the shore and a change of clothing. That evening, in a special service at the church, we were offered "the right hand of fellowship," and our membership was official....

Young people and children were highly valued. They were included in all aspects of the church's life and sat together in the front pews, sex segregated, not with their families. Our ministers reminded us constantly that we held the future of the church in our hands. The adults monitored our behavior for decency and orderliness, and expected us to tell the direct, simple truth at all times. Swearing was forbidden. Our work had solid economic value on the farm and "good worker" was a high compliment.

The members of our faith had historically been conscientious objectors [who refused to serve in combat in the armed forces], though I heard little discussion of it. During World War II, I knew several young men who were given alternative service in mental hospitals, and during the Korean Conflict, one served as a guinea pig to test the effects of irradiated food. Several joined the regular military, and I never heard a word of criticism of them for it. The Mennonites were loyal and hard-working Americans during World War II. They knew from whence they'd come and that the price of freedom often came high.

This group of Franciscan nuns arrived in North Dakota from Germany in 1928. "Teaching orders" of Catholic nuns, brothers, and priests established thousands of schools for Catholic children in the United States.

Ernte-Dank-Fest
Sept. 30. 1923

An Erntedankfest (harvest thanks festival) at a German American Protestant church in Massachusetts in 1923. Continuing a tradition from the old country, parishioners brought in samples of their garden and farm harvests. It was celebrated on September 30.

Children at the Fulton School in Cincinnati in 1908. As in public schools in other predominantly German communities, classes here were taught in the German language. Even non–German American children, like the girl at the left of the second row, had to learn German from kindergarten on.

SCHOOLS

In the 1850s, Henry Villard found a job as a schoolteacher in a Pennsylvania German community called Jonestown.

I duly entered upon my duties at the appointed time. Only 35 pupils out of 60 reported, and the attendance was always meagre, never exceeding 40, and sinking in bad weather often below 30. The ages of my pupils ranged from eighteen to five, the majority being perhaps twelve, with the sexes about equally divided. Most of them were healthy and comely, but shabbily dressed and anything but cleanly in appearance. The majority of them were evidently intelligent, but the examination with which I began my teaching...proved that they had received very irregular and limited teaching. Their spelling was very defective, their writing awkward, and their pronunciation of English very incorrect....

The general ignorance of my flock was amazing, and I seemed to be among veritable German peasant children. They saw very little of the outside world, and I readily attached them to me by reading and talking to them of it. Though they were the offspring of families that had been settled in Pennsylvania for generations, only a few of them could converse [in] or understand English, so I spoke German to them. At first they found it difficult to understand me, accustomed as they were to the dialect [spoken locally]. This very fact made them look upon the "schoolmaster," as they all called me, with awe, as a sort of superior being.

Herbert Heinen taught for nearly 40 years in the school in Comfort, Texas, in the Hill Country where many German immigrants settled. Heinen recalled his own school years in the same community during the 1880s.

A t that time the Comfort School was on the block now known as the park. The grounds were equipped with a gymnastics pole—*Turnstange*—a double bar and a long smooth ladder which were used for gymnastic purposes. A regular period of instruction in these gymnastics was set aside every Friday afternoon. The girls, of course, did not perform, but I recall at one time they were given instruction in marching and drills.

The old school bell was in a steepled tower on the roof. Two ropes hung down to be used by the teacher *only* for ringing that old school bell—20 or 30 peals before opening school and again at the close of classes.

Each teacher took care of two grades: the first and second in one room, the third and fourth in the other. There were no written tests, but a good deal of written work was done and preserved, with corrections, in copy books. One half of the lessons were supposed to be in English, the other half in German. Since German was...spoken on the school grounds and in the homes, it is understandable that very little English was learned.

Ernest L. Meyer recalled going with two friends to a gymnasium in Milwaukee in the early 20th century.

On Saturday morning I call for Hugo and Willie and we walk to the gymnasium at the West Side Turnerhall. Most boys of the neighborhood go to West Side or Bahnfrei, not because they are particularly athletic but because they are commanded to go. Our parents have made a kind of religion out of gymnastics, though we consider the thing rather boring....

But we would not dare cross our parents in anything; so we walk to the Turnerhall and on the way stop at the pretzel bakery. We ask for broken pretzels, and for a penny get a big sackful....

At the gym we put on gray shorts and blouses and gym shoes. There are about thirty boys in our class, and we are lucky to have Papa Brosius as a teacher. Papa Brosius is an institution, venerable and upright as an old cathedral. He must be sixty, and it looks as if he'll go on forever. He has snow-white hair, ruddy cheeks, and blue eyes, always happy. He is a bit corpulent, but stright as a pine and agile as a stripling. He puts us all to shame with his nimbleness on the trapeze, the bars, and the bucks. We love Papa Brosius, even though he works us hard. For when we falter he says nothing bitter; he merely stands before the class, performing the exercise vigorously, correctly, and out of sheer humility we fling ourselves into our work. We sweat, we heave, we stumble, and always this amazing old man flits in front of us, smiling and inexhaustible.

A German textbook for beginning readers, published in Cincinnati. The Gothic letters, most of which are the same as those in the Latin alphabet but are shaped differently, were standard in German American books and newspapers.

German Americans brought many educational innovations to the New World. Among them was the kindergarten, intended to prepare children for regular schooling. There are different accounts of who founded the first American kindergarten, but the honor is always given to a German American woman.

"You have much to learn, my little children, but remember it is the spirit as well as the muscles that counts," he says to the class.

"When Papa Brosius dies," whispers Hugo, "things won't be the same around here."

"He will never die," says Willie positively, almost defiantly, as if he were challenging something.

Edith Schallenberger was born in 1924 in Kirchhain, Germany. She and her mother came to Philadelphia in 1930 to join her father, working as a machinist. She remembered going to a German-language school.

Every Saturday I went to German School. It was at Turner's Hall, which was later taken over by Temple University. First it was gym classes and then swimming, and then in the afternoon we had German reading and writing classes. I learned German script. We also learned German songs.

Every so often, like prior to Christmas or Easter or whatever the holiday might be, they'd have like a *kinderfest* (children's party). The children would perform and the parents would come. Afterwards there would be dancing and food.

The German school was also connected with a German radio station. Maybe once a month the whole kindercorps would go and sing German songs over the radio station.

Alfred H. Kleine-Kreutzmann, who emigrated to the United States as a boy after World War II, described how he learned American English.

My parents and I came to this country in 1952. We came straight to Cincinnati because we had relatives here. I was almost twelve years old. Despite the fact that it was only six weeks to summer vacation, I was immediately sent to Catholic grade school and put into the fourth grade even though one year of "English" English in Germany had only prepared me to communicate such useful things as, "My Aunt Edith has chill blains."

The first thing the nun did was to take me to the public library and sign me up for a library card. I was a voracious

A Mennonite school in Hinkletown, Pennsylvania, in 1942. Some German American religious groups, such as the Amish and the Hutterites, still conduct classes in German today.

A sixth-grade class at a Baltimore public school where one-third of the classes were conducted in German. In Baltimore's high schools, art and music were taught only in German before 1917.

reader, and by the end of the summer I was reading adult books. I went to Dayton Street Branch and Grant Branch, both on the periphery of Over-the-Rhine [the old German neighborhood of Cincinnati]. Both had gotten rid of their German collections at the beginning of the war, but German material was easily available from the Main Library.

There was no question of our not being assimilated into the English-speaking mainstream of society. By the end of the first summer, reading, radio, and the movies had improved my English sufficiently for me to go on to the fifth grade. I was being assimilated so much faster than my parents that they were concerned that I might lose my German. So in order to get my allowance, I had to copy one page of German text a day. Or rather fill one page. This may account for my rather large handwriting.

Mary Conrad Van Grinsven remembered her experiences in a one-room schoolhouse in Minnesota during the 1950s.

Special memories in my life are of grade school days in a little red school house. This was a one-room red-brick building on the side of Grand Lake near Cold Spring, Minnesota. Grade school (way back then!) was the first grade through eighth grade with only one teacher.

An average number of students in the school for one year was sixteen. In my own grade we numbered three. Dan Voigt and Gary Court were my only classmates for all eight years!

For first grade I had Mrs. Margaret Rohloff. I recall her as being strict. Kids used to tease, "Roll off the cliff!..."

At nine o'clock school-day mornings, the brass bell clanged musically when Teacher would ring the hand bell.... Before any classes started, Teacher would always read to us for about ten minutes. This allowed us time to listen, rest, and also draw or color.... Subjects were taught at a large table in the front of the room, beginning with first grade, then second grade, etc. This was an opportunity to see and hear a brother, sister, or friend in their learning experience, though it did not necessarily interfere with our own personal studying....

In the back of the room sat another large work or activity table. In another corner was a huge oil furnace. It had a water reservoir with a faucet from its side so we had some warm water. Next to the furnace was the washing and drinking area. In a large sink-cabinet, we used two white enamel-ware bowls for hand-washing. A "red Wing" stoneware drinking water-fountain was also on the shelf.... Water was brought in every morning from an outside standing water-pump. Two stronger students would pump enough water to fill two large cream cans for daily use.... There was no indoor plumbing. No one really complained but we had to use out-houses!...

For many years the only playground unit was a merry-go-round. We could get it spinning so fast, a person could "fly" by hanging on tight with both hands and letting one's body whirl outward. A daring student could then "let go" and hope to miss crashing into the nearby steel flag-pole!

Carl Schurz

The '48ers—Germans who fled their homeland after the failure of the rebellions of 1848—were among the best educated immigrants ever to arrive in the United States. Among them was Carl Schurz, who made many notable contributions to his adopted country.

Schurz, then a fiery 19-year-old student, took part in rebellions in the Rhineland, the Palatinate, and Baden. He fled to Switzerland and then made a daring trip back to Germany to help a friend escape from jail.

Schurz found refuge in London, where he worked as a newspaper correspondent. In early 1852, he met Margaretha Meyer, from Hamburg, who was visiting relatives. In July 1852, the couple was married and a few months later left for the United States.

Eventually, the Schurzes settled in Wisconsin, where Carl became active in the antislavery movement. He gave speeches to German immigrants, urging them to join the newborn Republican party. In 1856, Margaretha opened a kindergarten, often said to be the first in the United States.

Carl Schurz led the Wisconsin delegation to the Republican convention in 1860 and threw the state's support behind Abraham Lincoln. As President, Lincoln rewarded Schurz by naming him ambassador to Spain. But Schurz hated the "elegant ease" of his life, while the United States was engaged in the Civil War. He returned home to become a brigadier general in the Union army.

After the war, Schurz became part owner of a German-language newspaper in St. Louis. He was elected senator from Missouri in 1869. Because he opposed the policies of Republican President Ulysses S. Grant, Schurz helped to form a short-lived third party. He then left politics to become a writer and journalist.

In 1877, President Rutherford B. Hayes named Schurz secretary of the interior. In this post, he advocated fair treatment for Native Americans, worked to protect forests on public lands, and introduced the first examination system for civil service jobs.

After leaving government in 1884, Schurz became editor of the *New York Post,* continuing his support for progressive causes. He died in New York City in 1906.

THE WORLD WARS

In 1914, the outbreak of World War I in Europe split many German American families, even though the United States did not enter the war until 1917. Hermann Hagedorn, born in Brooklyn of immigrant parents, recalled the conflicts within his family. Addie, his brother, had gone to school in Germany but lived in the United States. Their father, who by this time had retired and returned to Germany, wrote to Hermann.

It was inconceivable to Father that Addie or I in America might have any view of the life and death struggle other than his own. "Do everything you can to spread the truth about the war," he wrote me.

In case I might have forgotten what that "truth" was, he stated it again, point by point: "We [the Germans] were shamefully attacked. Everything had been prepared by our foes, years in advance, everything settled, signed and sealed. Britain is responsible for the whole thing; her envy, mistrust and jealousy were the cause. We had prospered too much."

Father's "truth" found me, in November, as unconvinced as I had been in August, when the Kaiser, in his war speech from the palace balcony, had set the official "line."...

[But] my emotions stood out against my reason. Soberly gratified though I might be at every German setback, every German victory set my Teutonic heart beating a little faster.... It made for tension and a feeling of guilt....

My beloved Addie, eight years my senior, whose brotherly counsel I had always found sound and had generally followed, was seeing the picture in other terms than mine....

He was convinced...that American financial interests, economically linked to Britain, were affecting American opinion and public policy, and that I myself was unduly influenced...by what I read in the papers. On almost every issue that arose between the Allied and the German point of view, Addie was inclined to lean toward the German, and I toward that of the Allies. The ever-sharpening tension between us was kept from the breaking point only by the genuine, stubborn affection we had for each other.

As I look back to those dark days, I feel ashamed that I never gave my brother credit for his courage in maintaining, throughout the period of American neutrality, a point of view, shared indeed, by millions of Americans in the Middle West and Far West but sharply out of key with the opinions of most of his friends and associates in business and in social life in the East. It required no courage on my part to take the position I took. I was flowing with the tide. I was everybody's "fair-haired boy," who was linked by blood, tradition and kin

During World War I, all German Americans were under suspicion of disloyalty to the United States. These farmers in Marion County, Kansas, display a flag as "proof" that they were good Americans.

to Germany, yet took the Allied side. But Addie stood almost alone, subject to malicious tongues and lifted eyebrows; and he stood fast.

After the United States entered World War I against Germany, German Americans suffered from suspicion and prejudice. Helen Wagner recalled the war years in the German American neighborhood of Yorkville, in New York City.

Those war years were really pathetic. You couldn't walk the street with a German paper under your arm. You'd be abused from one end of the block to the other. They went so far they abused the poor little German dogs that walked the street. That's the hatred that was. We kept speaking German at home, but we avoided it on the street. We had cousins and uncles over there [in the American army]. Lord knows how many of them were nearly killed by my brothers [who remained in Germany and served in the German army].

Elsie Mohr's family moved to Minnesota from Nebraska in 1902 when she was eleven. Most of her neighbors were Germans and spoke German at home. In an interview in 1974, she remembered the harassment that German Americans endured during World War I.

You know, there was an awful lot of hard feelings, you know, [by] those people that weren't German. But a lot of those who were German, you know, they wouldn't dare to talk about it.... You see, we didn't have nothing but German services in our church over here and mostly everyplace. And then when this got real bad, then they kind of abandoned that. They had to switch over more and more to English, you know. That's what they started then. That's the way it went because we couldn't have too much German service then.... There was a house then...I knew the people real well and their house was painted yellow [by vandals to stigmatize people, often without proof, as German sympathizers].... It was kind of rough alright.... We had to

The anti-German hysteria of World War I went to ridiculous lengths. Here, the statue of the mythical goddess Germania is removed from the Old Germania Life Insurance Building in St. Paul, Minnesota.

The kaiser, Emperor William II of Germany, was demonized by superpatriotic Americans during World War I. Here, a rowdy crowd in West Salem, Wisconsin, hangs an effigy of "Kaiser Bill" from a house where German Americans lived.

105

Groups such as the Turners retained interest in German affairs in spite of the anti-German sentiment during World War I and afterward. During the 1920s, a women's Turner club in Indianapolis helped German children orphaned during the war to come to the United States.

During World War II, German Americans wholeheartedly backed U.S. efforts to defeat Nazi Germany, and there was far less anti-German prejudice among other Americans. Here, German American women in Texas volunteer for Red Cross activities.

buy [war bonds]. We was kind of forced to, because you know if you didn't you then was yellow, you know.

During World War I, German Americans who could not produce citizenship papers were accused of being "enemy aliens" and often jailed. Oscar Ameringer, then editor of a German-language newspaper in Milwaukee, described the reasons why so many German immigrants in Wisconsin had failed to obtain citizenship papers.

Until shortly before the World War, very few immigrants took out naturalization papers in Wisconsin. When the Latin farmers [German refugees from the 1848 revolutions] settled in the wilds of Wisconsin there was nobody else around except Menominee Indians. They, as native Americans, would have been the proper people to issue naturalization papers to later arrivals. But somehow the Indians never got into the habit of issuing naturalization papers, whereupon the Forty-eighters simply "naturalized" themselves. There was nothing else for them to do if they wanted to organize their township and county governments. So they just voted and let it go at that....

[During the war] the local saviors of democracy would ask one of these old-timers: "Where were you born, Mr. Laubenheimer?"

"I was born in Germany."

"Aha, Germany! And how long have you lived in this country?"

"I've lived"—it might be sixty or seventy—"years in this country."

"Oh, that long? And when did you take out naturalization papers?"

"Naturalization papers? I never thought of taking out naturalization papers...."

"What? Why, you are an alien enemy and an illegal voter to boot! Come along, we'll teach you what true Americanism is."

When the United States entered World War II in 1941, Germany again became its enemy. General H. Norman Schwarzkopf spent his youth in Lawrenceville, New Jersey, before and during the war. His father, who had grown up in a German-speaking household, was a graduate of West Point. In his autobiography, It Doesn't Take a Hero, *General Schwarzkopf remembered the war years.*

When Pop was called back into the Army in 1940...sometimes he brought me to visit his encampment.... He took me to the room where he slept and showed me a book on his bedside table. It was *Mein Kampf* [*My Struggle,* a book by Adolf Hitler that described his plans for Germany]. He told me Hitler was a bad man and that a great deal of what he had written in that book was now happening in the world.... That was November 1941.

A few weeks later came the attack on Pearl Harbor, and it

finally dawned on me that something momentous was happening. When the news hit, I was sitting in a tree at Johnny Chiver's house. His mother told me, "This is terrible. It means your father will have to go to war!" I didn't understand what she meant—he was a soldier and away from home already. In the days that followed I listened to the news reports with my mother and sisters, but things didn't get any clearer....

Soon Lawrenceville was full of commotion. All the families had to put blackout shades on their windows, and the town ran air raid drills. When the air raid siren went off, that was the signal to pull down your shades.... Once we had a near panic because a blimp at Lakehurst Naval Air Station detected a German submarine off the coast. We really thought the Germans were about to invade New Jersey. We kept our blackout shades down continually for about a week.

By then, we boys were all armed to the teeth with toy machine guns and dressed in soldier suits and camouflage helmets, and we fought the Germans and the Japanese on a daily basis in our backyards. At school, the worst thing you could call somebody was a Nazi or a Jap. *Dummkopf* [dunderhead] became a popular insult, and a few times at school I had to explain that "Schwarzkopf" had nothing to do with *Dummkopf.* These incidents rarely led to blows because I was big for my age.

Ludwig Hofmeister came to the United States in 1925 and served in the U.S. Army during World War II. He remembered the onset of the war.

June 1941 I was drafted and went to Fort Dix [New Jersey].... I was in the army, and I forgot about everything else. I can honestly say I was very happy.... People ask me, "Did it ever come to you as a German to have to fight Germans?" Well, this feeling certainly is there. You can't deny it. But somehow I had so many things to be happy about in this country. I liked this country, and I owed certain things—because at twenty-one I had my own car already. Would I ever have a car in Germany? Never. I've been thinking about this in the later years, and I don't know...what I would have done if I ever would have faced a German. I think my impulse would have been to say something in German. I don't know.

I went to school in Providence, Rhode Island, to the Brown University, for [training] interpreters. This was in 1942. When we went overseas, my job was interpreter of prisoners of war. After the invasion [of Europe] in 1944, I had to be an interpreter in Germany, in the Rhineland, and I only had to deal with civilians....

There was one big Nazi, and it was a pleasure for me to arrest him. He must have been one of them guys who made himself rich, because he was living very, very, very luxurious. I had no pity on him, because he was a big fat guy, and I didn't like him.

Dwight D. Eisenhower

"I come from the heart of America," Dwight Eisenhower said in June 1945, speaking at a victory ceremony in London. Less than a month before, Eisenhower had accepted the surrender of the German armies, ending World War II in Europe.

Eisenhower was born in Texas, one of seven sons of David and Ida Eisenhower, and he grew up in Abilene, Kansas. The Eisenhowers were members of the River Brethren, a Mennonite sect that opposed all forms of violence, including warfare. They were poor, and although Dwight did well in school, it seemed unlikely that he could afford college.

An opportunity arose when Dwight received the second highest score in a state exam for applicants to the U.S. Military Academy at West Point. The highest scorer failed the physical exam, and Dwight received the appointment. His mother wept when he left home, feeling that a military career was a disgrace.

Eisenhower graduated from West Point too late to see service in World War I. For the next 25 years he held a variety of posts in the peacetime army. His keen mind impressed his superiors, however, and when the United States entered World War II in 1941, Eisenhower was called to Washington, D.C. There, he served on the military staff that planned war operations. In 1943, he was asked to write a plan for the invasion of Nazi-held Europe. He was then given the job of commanding the Allied forces that would carry it out.

Eisenhower was a brilliant leader whose greatest talent was inspiring those who served under him. The troops referred to him by his lifelong nickname, Ike, which later became part of one of the most famous Presidential campaign slogans in U.S. history: "I like Ike." He easily won the 1952 election for President and won a second term four years later.

President Eisenhower guided the nation through a difficult era, when tensions between the United States and the Soviet Union threatened to erupt in nuclear war. In his farewell speech, however, he warned against the danger of a "military-industrial complex" that would absorb too much of the nation's resources in preparing for war. It was a message that reflected his roots.

A German American family dresses up for a Steuben Day parade in Chicago. Friedrich von Steuben was a Prussian military officer who trained George Washington's troops during the Revolutionary War.

CHAPTER SIX

PART OF AMERICA

believe the children of every race have brought with them something worthy of reception and absorption by the American people," wrote Richard Bartholdt, a German immigrant who became a U.S. congressman in the early 20th century. He added, "America cannot be made Anglo-Saxon any more than it can be made Irish, German, French or Italian."

German American influences are deeply embedded in the culture of the United States, so much so that many people are unaware of their origins. Two of the United States's most popular foods, hamburgers and hot dogs, are German American contributions. Apple strudel, sauerkraut, potato salad, and mustard are among the other American foods of German origin.

The first U.S. kindergartens ("children's gardens" in German) were started by German '48ers. The gymnasiums and physical education programs in virtually every U.S. high school reflect the influences of the German American Turners who inaugurated such programs of physical fitness in the 19th century.

Most aspects of the Christmas holiday as celebrated in the United States come from the cus-

toms of German immigrants. The Germans who settled in Pennsylvania in the 18th century held gala celebrations on Christmas Eve and Christmas Day. They gave each other gifts and toasted the holiday with feasting, beer, and song. The English neighbors of those early Germans, whether they were Puritans, Quakers, Baptists, or Presbyterians, disapproved of such festivities, regarding them as pagan traditions.

In fact, the use of a Christmas tree (which first appeared in the United States in Lancaster, Pennsylvania, in 1821) began as a German custom in pre-Christian times. An evergreen tree, still colorful in the dead of winter, was a reminder that spring would someday come. Pennsylvania Germans decorated their trees with candy, fruit, and gingerbread cut into the shapes of stars, sheep, hearts, and houses.

The German-style celebration of the birth of the *Christkindl*, or Christ child, gradually became accepted by other Christians in the United States. Over time, *Christkindl* came to be pronounced Kris Kringle, another name for Santa Claus, who comes from yet another German American tradition.

On December 6, Germans cel-

ebrated the feast of St. Nicholas, a Christian bishop who lived in the fourth century. St. Nicholas, called Sinterklaas by Pennsylvania Germans, was known for his generosity, particularly to children. The custom of giving gifts to children on this saint's feast day spread from Europe to America with German immigrants. Frequently, men put on false beards and bishops' hats to distribute the gifts.

In the 1880s, the German American cartoonist Thomas Nast first drew the image of the fat, jolly, white-bearded man in red who is today's Santa Claus. In this form St. Nicholas became jolly old St. Nick, but originally he was sometimes a scary figure. Children were warned that if they were not good, St. Nicholas would punish them. Darinda Schmidt recalled that when she was a girl in Idaho at the beginning of the 20th century, "Saint Nicholas and his helpers...stormed into homes while young children clung to their mother's knees in fright. It was not uncommon for children, that had misbehaved or did not know their prayers, to be stuffed into burlap sacks and carted out of the family home, driven for what seemed like miles, and then deposited at the end of the family driveway."

In Fredericksburg, Texas, settled by German immigrants in the 1840s, people still light fires in the fields the night before Easter, telling children that the Easter rabbit is making dye for the eggs by burning wildflowers.

The Easter bunny and Easter eggs also originated as German American customs. The hare, a larger relative of the rabbit, is born with its eyes open and thus in some ancient religions became identified with the full moon. Because the Christian holiday of Easter is celebrated on the Sunday after the first full moon of spring, the hare became a symbol of Easter. Pennsylvania German children built nests in the fields for the Easter hare, believing that if they were good the hare would lay eggs in the nests. On Easter, the children went into the fields with baskets to collect the eggs, which had of course been set out by their parents.

The custom of coloring eggs began in Europe during the Middle Ages. Eggs were a special Easter treat in those times, for Christians were forbidden to eat them during the penitential season of Lent. On Easter the parish priest often distributed eggs, with their shells dyed red to symbolize the blood of Christ. In Fredericksburg, Texas, settled by German immigrants in the 1840s, people still light fires in the fields the night before Easter, telling children that the Easter rabbit is making dye for the eggs by burning wildflowers.

The German American community tended to lose its strong ethnic character after the two

world wars in which Germany was an enemy of the United States. Some German-language newspapers still survive, such as the *New Yorker Staats Zeitung und Herold,* which is now more than 150 years old. But the old German American neighborhoods in the major cities have virtually disappeared. Along East 86th Street in Manhattan, where every store once had a German name, there remain only a few German

A float in an Oktoberfest parade in LaCrosse, Wisconsin, in 1968.

restaurants, candy shops, and specialty food stores. Many of the customers tend to be visitors from the suburbs who retain their fondness for the foods they ate as children.

Even so, German Americans have never ceased to play a major role in all facets of life in the United States. An immigrant named Charles Steinmetz, who was nearly turned away at Ellis Island in 1889 because of a physi-

cal deformity, went to work for the General Electric Company, where he revolutionized the science of electrical engineering. His work in large part made possible the huge electric generators that today send power to virtually every American home. The brand names of many everyday products likewise reflect their German American origins: H. J. Heinz canned foods, Mueller's noodles, Hecker flour, Folger's coffee—and of course beers such as Schlitz, Miller, Pabst, Stroh's, and Budweiser. John J. Bausch, an optician and lens grinder, and Henry Lomb, whose names signify quality on modern optical products from binoculars to microscopes, were German immigrants. George Westinghouse, born to immigrant parents in 1846, founded the company that bears his name and today is involved in diverse activities, from broadcasting to manufacturing electrical appliances.

Few think of George Herman "Babe" Ruth as a German American, but his parents were immigrants whose name was Erhardt. From Casey Stengel to Roger Staubach, Wes Unseld, Mike Schmidt, and Bret Saberhagen, many other German Americans have been major sports stars.

The literature of the United States has also been enriched by the works of such German Americans as Theodore Dreiser, H. L.

> As the memories of war faded in the 1950s, the German American community began to regroup by holding German Days and Steuben Day parades on September 17, the birthday of the Prussian officer who trained George Washington's army.

Mencken, John Steinbeck, Sylvia Plath, and Kurt Vonnegut.

The United States greatly benefited from the talents of those who fled Hitler's Nazi regime in the 1930s. In addition to Albert Einstein and other scientists, these refugees included titans of the music world such as Otto Klemperer, Erich Leinsdorf, Arnold Schoenberg, and Artur Schnabel. Walter Gropius and Ludwig Mies van der Rohe helped to create the distinctive glass-walled skyscrapers of modern architecture. Photographer Alfred Eisenstadt, movie director Billy Wilder, theologian Paul Tillich, and psychologist Bruno Bettelheim all put their talents to use in the United States.

Lawrence Welk, born in the German-speaking community of Strasburg, North Dakota, became one of the biggest stars of the early years of television. In the 1950s, when rock and roll became popular, Welk's musical variety show attracted huge audiences with its old-fashioned music and songs that Welk had once played at German American wedding parties.

As the memories of war faded in the 1950s, the German American community began to regroup by holding German Days and Steuben Day parades on September 17, the birthday of the Prussian officer who trained George Washington's army. Many German American clubs have revived old traditions

Fritz Felton of Hays, Kansas, works on a memorial to the Volga River Germans who first settled in Kansas in 1876.

such as the *Schlachtfest*, when pigs are brought to butcher while a band plays a funeral march.

Sängerbund, or singing societies, keep alive the German tradition of choral music. And *Sängerfeste*, like Cincinnati's Oktoberfest, are gala occasions when clubs gather to display their talents. Many Turnvereine still carry on their gymnastic activities and today often field soccer teams as well.

German Americans still celebrate many holidays that remain distinctively German. *Fasching* is the German pre-Lenten feast, celebrated with as much gusto as the equivalent holiday of Mardi Gras.

Oktoberfest is just what the name implies—an autumn harvest festival where hearty German food and beer are served and consumed. Cincinnati claims to have the largest Oktoberfest in the United States, but it is rivaled by a festival held in LaCrosse, Wisconsin.

New Braunfels, Texas, holds a Wurstfest each November, featuring music, dance, and many kinds of sausages that make the ordinary wiener seem tame. Men dress in *lederhosen* (short leather pants) and the women wear *dirndls* (colorful, full-skirted jumpers) to celebrate their German Texan heritage.

In 1983, German Americans observed the 300th anniversary of the settlement of Germantown, Pennsylvania. The U.S. government issued a stamp to mark the occasion. All over the nation, with speeches, music, and public ceremonies, German Americans demonstrated their pride in a three-centuries-long record of achievement.

I love the German language. She is the real *Heimat* [homeland]; all others are adopted children who, at first glance, can easily be confused with one's own flesh and blood. Languages are beautiful bridges. They lead from one shore to the other.

And America? I love America the way an orphan loved his kindhearted foster parents and the first room of his own. Without reservations.

—Sabine Reichel, who was born in Hamburg, Germany, in 1946, and emigrated in 1976

Michael Miller, a descendant of Germans from the Black Sea region, returned to the Ukraine in 1994 to search for his roots. Here, he helps a villager draw water from a well.

GERMAN AMERICANS TODAY

The German American community became less visible after World War II. In recent years, however, there has been a revival of interest in the best parts of their heritage. The novelist Kurt Vonnegut—whose German ancestors, he has said, came to the United States "long before the Statue of Liberty was built"— expressed his feelings in 1991, when East Germany was politically reunited with West Germany.

People ask me how I feel about German reunification, and I reply that most of what we like about German culture came from many Germanys. What we have good reason to hate about it has come from one....

The hatred for all things German expressed by Anglos in this country during World War I (before my birth) was so virulent that there were virtually no proudly German institutions still operating (I include my father) when it came time for World War II. German-Americans had become (in self-defense and in embarrassment over Kaiser Wilhelm and then Hitler) the least tribal and most acculturated segment of our white population. (Who was Goethe? Who was Schiller? Ask Casey Stengel or Dwight David Eisenhower.)

One American in four is descended from German immigrants, but what politician nowadays ponders how to woo the German vote? (That is OK with me.)

Many new German immigrants came to the United States in the years after World War II. Alfred H. Kleine-Kreutzmann, who arrived with his parents in 1952, lived in the Over-the-Rhine neighborhood of Cincinnati, where German immigrants had settled since the mid-19th century. He describes the changes in that area since the 1950s.

In those days one could still get all the necessities of life in the Over-the-Rhine area by speaking German, especially at Findlay Market and in a large number of specialty shops on Main Street above Twelfth. The shopkeepers could understand German even if they did not speak it. The small shops had passed out of existence by the beginning of the sixties. Findlay Market is still there and still carries German specialties.... Of the eating places for which Over-the-Rhine had been famous, only two remained by the fifties: Grammer's, a restaurant, and Stenger's, a saloon. Grammer's eventually diluted its German traditions and doesn't exist any more. Stenger's has changed very little and flourishes....

As far as I am concerned, Cincinnati's German heritage lives on most satisfactorily in its academic institutions, its musical organizations, and the cultural opportunities of the Public Library.

Michael M. Miller is the bibliographer for the Germans from Russia Heritage Collection at the North Dakota State University Libraries. He told an interviewer about his 1994 trip to Odessa, Ukraine, the home of his ancestors.

For me it was like going back in time. I had seen it in pictures, but I didn't expect that it was still that primitive in the former German villages. My grandparents on my father's side came from the Catholic German village of Krasna in Bessarabia in today's Moldavia. My mother's family, the Baumgartners, came from Strassburg in today's Ukraine.

I had hoped to find cemeteries where my ancestors were buried, but the former Communist government tried to get rid of the traces of the German society that had existed. But what touched me was the large, beautifully designed churches that these Germans, with few resources, came together to build. Many are similar to cathedrals. I've often wondered if the tradition came from Germany. There is St. Michael's Basilica in Selz, Ukraine. Lawrence Welk's parents came from there. The churches were symbols of their strong belief in Christianity. And later the immigrants built similar churches in North Dakota.

In Ukraine, I saw the same kinds of thick-walled sod and clay brick houses that I remember growing up in south-central North Dakota. I had hoped to find the houses where my ancestors lived, but it was difficult to locate them. Generally speaking, the German Russians, when they came to North Dakota, didn't want to talk at all about the old country, because the memories were painful.

In 1991, the German-Russians living in the former Soviet Union learned about the Germans from Russia in North Dakota from an article in a German-language newspaper, *Neues Leben,* which is published in Moscow. Since then, Russian Germans have been writing letters to us at the Germans from Russia Heritage Collection, and we have been able to put people in touch with relatives whose ancestors immigrated to America.

One of the most interesting people we heard from was Paul Krüger, who formerly lived in Siberia. We discovered that his uncle Otto Krueger was a United States Congressman from North Dakota in the 1950s. When Otto left for the United States in 1910, he left behind a 13-year-old brother, Bernhardt. During World War II, Bernhardt was sent to a labor camp in central Asia. He was still there while his brother was serving in Congress. Paul, his son, lived there too. It is only now that the descendants on both sides of the family are learning what happened to their relatives.

The Germans from Russia Heritage Collection is trying to interview as many people as possible in both countries to put together family histories. And we're organizing tours for German-Russian Americans to their ancestral villages in southern Ukraine.

Dennis Franz

In 1994, two of the stars of "NYPD Blue" were nominated for an Emmy as best actor in a TV dramatic series. One was handsome, slim, red-haired David Caruso, who was regarded as the leading man of the hit new detective show. The other was pudgy, bald, middle-aged (49) Dennis Franz, who portrayed Caruso's gruff partner.

The audience exploded in cheers as Franz was awarded the Emmy. It was a victory for "average" guys everywhere and a tribute to Franz's acting ability, honed over 25 years in show business.

Born in a suburb of Chicago, Franz is the son of German immigrants Franz and Eleanor Schlachta, both postal workers. In high school, Dennis's girlfriend talked him into going to tryouts for the school play. "The guys trying out were sort of meek," Franz recalls. "I used to sing when I was a kid, and my father's words to me were, 'Be loud.' So I got up and was loud. And I got one of the leads."

It was several years, however, before he began to pursue a professional acting career. After attending Southern Illinois University, Franz was drafted into the army and sent to Vietnam. He has a hard time recalling his experiences in combat, he says. "I chose to let a lot of it go. You spent every day trying to survive."

After being discharged in 1970, he returned to Chicago and joined a theater group. From the first, he was offered roles as a detective and police officer. After bit roles in two movies, he moved to Los Angeles. He got his first big break in 1982, when TV producer Steven Bochco cast him in the police series "Hill Street Blues." Twelve years later, Franz was Bochco's first choice for the role of Lieutenant Andy Sipowicz in "NYPD Blue."

According to Franz, he has played 28 different police roles in his career. He is not particularly happy about this. "An actor can be anybody," he says. The intensity he conveys as Sipowicz, whose rage at the evils he deals with is barely controlled, suggests that bigger roles may be in store for him.

PRESERVING THE HERITAGE

The queen and her attendants at a Bavarian festival in Michigan. Many German Americans still retain their ties to a particular region of Germany, with its own distinct customs and traditions.

The German mode of celebrating Christmas was adopted by most other Christians in the United States. Hermann Hagedorn remembered the Christmases of his childhood in the late 19th century in Brooklyn, New York.

Christmas began for us about the first of December, when Father and Mother would suggest that it might be prudent for us youngsters to make a list of things hoped for, to be a guide to the busy *Weihnachtsmann* [Christmas man]. Was that a moment!

A week after the making of the *Weihnachtsliste*, Mother would casually remark some morning that she was going to start preparing the dough for the *Honigkuchen* [honey cookies]. Those golden-brown cookies were an essential part of the show. You couldn't imagine Christmas without them. Mother made them by the barrel and though she gave them away generously, we generally had some until well into February, even with three hungry children reaching into the deep tins....

We might help bake the cookies, but, when the pans went into the oven, that was the last we saw of the *Honigkuchen* until Christmas. There was no such thing as anticipating the occasion. This particular kind of cookie was a part of Christmas, never made at any other time of the year and never, under any circumstances, eaten in advance of the day.

The day, of course, by German tradition, was not Christmas Day itself, but Christmas Eve.... The glittering splendor of the dinner table of those far-off Christmas Eves! The suspense of that final hour before the giving of the gifts, the *Bescheerung!* Just before the dessert, Father would excuse himself. "Come up in about half an hour," he would say. We knew that he would be bringing Mother's presents to her table in the Christmas room, the *Weihnachtsstube*, and then lighting the tree.

Long before he was ready for us, we would be in the corridor outside the parlor, sitting on the stairs. Mother would begin to sing and we would all join in, even to Grandpa....

I can still hear Father's little silver bell, and when I can't hear anything else, I think I shall still hear it. For through the tinkle, I would become aware of a dazzling scene, as Father, beaming, opened the door; every gaslight on the crystal chandelier blazing, a Christmas tree sparkling with a hundred living flames and, roundabout, on all sides, a splendor of gifts.

Each of us had his own table and, of course, made a dash for it. Most of the gifts would be useful, the sort of thing Mother would have had to buy for us in the course of the winter anyway, but, finding it there, under the blaze of lights, gave

it an aura. And then there were always things that were not utilitarian, a sled, perhaps, and lead soldiers and stamps and books; and on every table, always a soup plate full of the *Honigkuchen* [honey cookies] and another full of nuts and raisins and tough jaw-cracking *Pfeffernüsse* [spicy cookies] and *Aniseplätzchen* [anise-flavored cookies].

I remember the relaxing of the tensions, sitting on the floor with a new toy or a new book, and looking up at the tree. A more cautious generation, stringing hard-faced little electric bulbs up and down and across its Christmas trees, knows nothing of the enchantment that was in those wavering little yellow flames, making the tinsel sparkle and the gold cheeks of the walnuts glow fitfully....

Father hovered over it all, warm and tender and, I am sure, completely happy.

The Easter rabbit was another custom brought to America by German immigrants. Gilbert J. Jordan described the Easter festivities in his Texas community.

Like many of our Christmas customs, the Easter rabbit and the colored eggs came from Germany. However, our parents were smart enough not to try to sell us on the idea of an egg-laying rabbit. Even the kids knew too much about rabbits and birds to accept this folklore, no matter how German it might be. Our Easter rabbit had a much easier time of it. He simply dyed the hard-boiled eggs and put them into flowery nests (*Osternester*) that we prepared for him. I don't remember how much faith the little kids had in the rabbit's activities, but among adults he was a questionable character, so they referred to an unreliable person as an *Osterhase* (Easter rabbit).

On the Saturday before Easter we children always went on a flower-hunting expedition. We took a child's wagon and some baskets or buckets and went out to the pasture and fields to gather flowers and fresh grass for our Easter nests. We fixed

the nests on the front porch, where the Easter rabbit could easily find them.... Naturally we children never had any part in the egg dyeing; this was the Easter rabbit's job. The day after Easter was "Second Easter" *(Zweite Ostern)* for us. Like "Second Christmas" this was a holiday, and most of the people spent the day visiting friends and relatives.

Wilfred "Bill" Eckhardt is a third-generation German American who has been a Lutheran campus minister at the University of Iowa for 25 years. He writes of his heritage.

My grandfathers helped me appreciate my German heritage. Each worked very hard, was frugal, enjoyed a good laugh, respected authority, and valued his vocation and education. Each was stubborn. None of my grandparents wasted anything and criticized anyone who did. They were tenderhearted. My mother's father, Ernest Kirchhof who farmed in Humboldt County, Iowa, gently sobbed when I asked about the three adjacent graves of young Kirchhof children in our family cemetery east of Thor, Iowa. Others had to tell me that they were his two young brothers and little sister who died of typhoid fever in a six-week period sixty years earlier. Each grandfather also lost children of his own. It was important to them for their children to succeed. Grandfather Fredrick Eckhardt, a sausage maker and owner of a general store in Boyd, Minnesota, generously helped my dad and his brothers get started in business and three of his four daughters get a college education. Religion was important to him, as is demonstrated by his being a charter member of both Zion Lutheran Church in Mora, Minnesota, and Zion

In Cincinnati, Oktoberfest is celebrated each year in the city's central business district. Signs rename the city "Zinzinnati," an affectionate reminder of the way German immigrants pronounced the name of their new home.

Lutheran Church in Tyrol township. His father was the contractor and he was the only carpenter for the $1,000 first structure of the Tyrol church.

My grandfathers were well informed by any standards. Grandpa Eckhardt could discuss theology, history, geography, gardening, and especially military strategy, his father having been a Prussian mercenary officer. He had been strongly criticized during World War I for being German. He followed events of World War II, not just because my father served in the American army, but because he was justifiably concerned about relatives in Germany, most of whom lived in the former East Germany. I remember him pondering the news with his ear near the radio and his chin in his hands. He often marked battle situations on a large color map of Europe which hung in the meat cutting room of his store.

My grandparents were successful and happy because they viewed life as an uncertain, hazardous but exciting journey. They responded to adversity by making the experience a building block for the future. Moreover each grandfather had an epic personal vision that formed an attitude of confidence and humility. Hope was based on an ethic of hard work, resiliency, fair and generous treatment of others and trust in a merciful God. Their God had rules of conduct which governed their human lives as surely as they governed nature. God had a plan that fit them into the world and made them know a personal significance. They understood the tyranny of a society where the chance for advancement was remote, so they valued freedom. Even more, they valued opportunity. America allowed them to use their personal energy to care for themselves, their family and to expand the horizons of each. They appreciated America for that gift. This incentive of appreciation was passed on to the following generations.

Life for my grandfathers was not so serious they couldn't enjoy a humorous story—unless it was on them. Grandfather Eckhardt's first "horseless carriage" was a new Chevrolet "Model 490," called that because it cost $490. Alfred, his oldest son, gave him a quick lesson in steering, speed control, and the use of the reverse pedal. My dad, the youngest son, was ten years old, and got to ride as Grandpa practiced driving in the pasture. Dad remembered that Grandpa never swore around the kids. They returned to put the prize away in the machine shed. They went down the hill and entered the shed's open door when Grandpa pulled on the steering wheel with both hands and excitedly shouted, "Whoa! Whoa!" The Chevrolet kept going through the back wall of the building as boards, straw, and chickens flew. They crashed into the tree behind the shed. Neither occupant was hurt, but the tree and the "490" certainly were. My dad says that Grandpa didn't use a swear word. He just repaired the car. The story wasn't repeated around him.

George Keilhofer, a wood-carver in Frankenmuth, Michigan, puts the finishing touches on a shield inscribed with the traditional German eagle.

THE MAIER FAMILY

Paul Maier's grandparents shortly after their arrival in Boston in 1880.

Paul L. Maier is a professor of ancient history at Western Michigan University in Kalamazoo and campus chaplain to Lutheran students. A graduate of Harvard University and Concordia Seminary in St. Louis, he pursued postgraduate studies at the Universities of Heidelberg, Germany, and Basel, Switzerland. Dr. Maier has published 13 books in 10 languages, with several million copies in print. His latest is a new translation from Greek of the works of the 1st-century Jewish historian Flavius Josephus.

Q: *Can you tell us about your German ancestors who came to the United States?*

The Germanic roots are deep on both sides of our family tree. On my father's side, my grandfather Maier came from Württemberg in southwestern Germany, and my grandmother from Rothenburg on the Tauber, the picturesque medieval town in northern Bavaria. After a perilous voyage across the Atlantic in November 1880, in which their small sailing vessel with auxiliary engine was almost lost at sea, they settled in Boston. On my mother's side, both grandparents were second-generation German Americans, and their names Eickhoff and Kuehn more than demonstrate their Teutonic heritage.

Grandmother Maier, whom we called "Grossie" (for *Grossmutter*), often visited us at Christmastime. Born in southern Germany, she never lost her German accent, and pronounced, for example, "faith" and "fate" exactly the same way. Before Christmas, Grossie filled the air with the fragrance of cookies she baked from wondrous old-world recipes. We were convinced that her *Prügelkrapfen* would be "The Official Cookie" in heaven.

Q: *Tell us about your parents and how they passed down their German heritage during your childhood.*

My father, Walter A. Maier, was a Harvard Ph.D. who was a magazine editor, Old Testament professor, author of 31 books, and probably best-known as the founding speaker of "The Lutheran Hour," a radio program he started in a South St. Louis attic, which grew to a worldwide broadcasting enterprise heard in 36 languages in 120 nations and territories.

My mother, Hulda A. Eickhoff, was an Indianapolis schoolteacher who married my father in 1924. Both my parents knew German very well, but, unlike *their* parents, they spoke English without any German accent. But since they had been raised in families where German was spoken before English, they tried the same with their own offspring, my brother Walter, Jr. and me, with somewhat mixed results.

One of my earliest recollections is a conversation I had with my mother—in German. When I was a toddler, she used to take me sight-seeing to various churches near our home in St. Louis County, but one day she varied her routine by showing me a synagogue. I looked inside, then asked, "Aber wo ist der liebe Heiland?" (Where is the dear Savior?) When Mother explained that you didn't look for Jesus in a Jewish synagogue, I replied, "Komm, lass't uns gehen." (Come, let's go.) This, however, never led to any prejudice in the Maier family, I'm glad to report.

As is typical of German, Mother's rare expletives in that language always sounded rather mild when translated. Once, in the early '30s, when the engine in our Hupmobile resolutely refused to start, I recall her hitting the steering wheel and shouting, "*Du fatale* beast!"—a bilingualism that means "You disagreeable beast!"

Brother Walt carried on with German somewhat longer than did I, but his bilingual talents halted abruptly the first time he played with neighboring boys and asked them to throw him a ball by saying, "Gib mir den Ball!"

"So who's *that* foreigner?" they sniffed.

He abandoned German in public after that, except in our family devotions, where we often prayed the Lord's Prayer in the "Vater unser..." mode. Our bedtime prayers also continued in German throughout our youth: a series of hymn verses and Bible passages. Years later, we all recited them together again at Father's bedside as he lay dying in January 1950.

Every Christmas the Maier family celebration started with Father at the piano playing (by ear) the German Yuletide folk song, "Kling, Glöckchen, Kling-a-ling-a-ling" ("Ring, Little Bell"). The rest of the family sang the lyrics lustily while clanging bells and chimes of every sort as we all marched in festive procession throughout the house during endless choruses.

Whenever the Maier family embarked on a vacation, Father would lead us in prayer for a safe journey. The invocation always began, "Unsern Ausgang segne Gott, unsern Eingang gleichermassen...." (Our departure may God bless, our return also....) Except for such prayers, English now became exclusive in our home life, although emergencies might still evoke the German in which Mother was educated. While driving, for example, if an accident seemed imminent, Mother would instantly cry out, "Gott behütte uns!" (God protect us!) We did have several accidents on those narrow, twisting, two-lane roads of the 1930s, and Walt and I grew to hate Mother's three-word shout, since it always seemed to herald imminent disaster!

Mother's three sisters in Indianapolis were never our "aunts" but "die Tanten," and whenever we visited them we always learned to close the screen door as quickly as possible,

PRÜGELKRAPFEN

This is Paul Maier's grandmother's recipe.

INGREDIENTS

1 lb. granulated sugar
1 lb. butter
4 eggs
2 tsp. cinnamon
1 lb. chopped almonds

Stir the above, adding 1 lb., 4 oz. flour—enough to roll flat (better a little too much than too little). Roll out into cookies, 1/8-inch thick. Bake quickly in well-heated 375 degree oven. Sprinkle a mix of confectioner's powdered sugar and cinnamon over the warm, baked cookies or roll them in the sugar mix.

Walter Maier and Hulda Eickhoff, Paul Maier's parents, on their wedding day, June 14, 1924.

Paul Maier, age four, looking at his father's German Luther Bible, printed in 1541.

The Maier family before a "Lutheran Hour" rally at the Chicago Stadium in 1943. From left, Paul, his grandmother, mother, father, and older brother, Walter, Jr.

or Lydia, the eldest, would cry, "Jammer, die Fliegen!" (O misery! The flies!) Whenever the aunts visited us at our summer home along Lake Ontario, morning table devotions often began with what Walt and I (now thoroughly anglicized) rather irreverently called "The Fang Song"—"Fang dein Werk mit Jesu an" (With the Lord begin your task). At day's end, our parents often stood arm in arm along the lakeshore and serenaded the setting sun with the song, "Seh't, wie die Sonne dort sinket...."

Q: Aside from your brother's experience, were there any other occasions when you felt prejudice because of your German background?

The blending of a foreign culture into the American mainstream was probably easier for German Americans than for most other ethnic groups because of the vast number of German immigrants to our shores. Never in my childhood during the 1930s did I feel any sort of "minority complex," although I will admit that my horizons were very well peopled with other German Americans, especially Lutherans.

My own church body began with the arrival of four boatloads of Saxon immigrants who docked at New Orleans and shipped up the Mississippi to St. Louis. In 1847, they founded the Deutsche evangelish-lutherische Synode von Missouri, Ohio, und anderen Staaten, today mercifully abbreviated to The Lutheran Church—Missouri Synod. For decades afterwards, German was used exclusively for worship. In fact, it was a common opinion that God himself used only German. As proof, Genesis 3:9 was quoted (from the German Bible, of course!), where God said, "Adam, wo bist du?" ("Adam, where are you?") In later years, English services were added, and finally English prevailed. A few of the great old Lutheran churches, however, *still* offer one of the Sunday services in German.

Historically, however, there would be an exception: the anti-German sentiments kindled during World War I. It is ironic to contrast how America reacted to German culture during the two world wars. In World War I, for which Germany was *far* less responsible, the antipathy, paradoxically, was much stronger. Sauerkraut was renamed "liberty cabbage," kindergarten was dropped in schools, German was no longer taught at universities, and no music by Bach, Beethoven, or Brahms was performed. In World War II, however, for which Hitler and Germany were *far more* responsible, America seemed to have matured enough not to attack German culture in the process.

In the mid '30s, Mother hosted special parties at our home in which only German was spoken by the guests. The German consul in St. Louis and his wife attended some of these, until the nefarious shadow of Adolf Hitler put an end to that.

Most German Americans had to go through an agonizing reappraisal regarding Hitler. In the earliest '30s, he was often regarded as something of a Teutonic hero, the man who built the autobahns, the Volkswagen, and made the trains run on time. While still a lad in the '30s, I recall a conversation I had with two older German American friends in Boston. One of them said: "The world must now realize that Adolf Hitler is either a madman or a genius. Now, it's clear that he's not insane, so he must be a genius." Several years later, they—and myriads like them—were singing a dramatically different tune!

Q: After World War II, your father saw firsthand the destruction Hitler had caused.

In 1947, my father was invited by the U.S. War Department to fly to Europe to advise the Allied Military Government for Germany on educational and religious matters. He had no trouble rehabilitating his German language skills on this trip, and he loved to tell of the time he was passing out candy, chocolates, cheese, and fruit from the PX [store on a military base] in Germany to some near-starving Germans who were destitute after the war. One woman took him aside and asked in all seriousness: "Please tell me: aren't you the angel Gabriel? Scripture says Gabriel will come to earth in human form and do much good. It must be you!" The pious woman was rational, but the situation had overwhelmed her. "No, my dear woman," he replied, then added a pun, again lost in translation: "I'm hardly an *Engel* (angel). It's more like a *Bengel* (rascal)!"

Q: Have you and your own children carried on the German cultural traditions?

I retrieved my German at the University of Heidelberg, and then transferred to the University of Basel in Switzerland, where I took my doctoral exams in German. I went on to marry a German American girl (as her name suggests: Joan M. Ludtke), even though I had set no ethnic limitations where romance was concerned! She and our daughters shared the Germanisms with Mother Maier at Christmastime, as well as the phrase "Gesegnete Mahlzeit" (blessed mealtime) after table prayer. Our girls also elected to take German in school, and Julie, our second daughter, spent her junior year studying at Reutlingen in Germany.

Although my mother was born and raised in Indiana, speaking and writing English perfectly, the German confirmation instructions she received and the German hymns she learned as a child stayed with her indelibly throughout life, especially toward its close. During the months before she passed away in 1986—on her 96th birthday—she would recite endless verses of German hymns from memory, finding extraordinary inspiration and comfort in them. Christianity and music, then, served as the principal expressions for German culture in our family. I doubt that our experience in this regard was unique.

Dr. Walter A. Maier, at the piano, accompanies the chorus from "The Lutheran Hour" radio program. His wife sits beside him.

Joan (second from left) and Paul Maier in 1995 with their four daughters (from left) Julie, Katie, Laura, and Krista.

GERMAN AMERICAN TIMELINE

1683
Led by Franz Daniel Pastorius, 13 families from the town of Krefeld arrive in Philadelphia Harbor on the ship *Concord*.

1688
Pastorius writes the first North American protest against the institution of slavery.

1732
Benjamin Franklin publishes the first German-language newspaper in America—the *Philadelphische Zeitung* (*Philadelphia Times*).

1735
John Peter Zenger, publisher of the *New York Weekly Journal,* is acquitted of the charge of libel—a victory for freedom of the press.

1741
Hans Nicholas Eisenhaur, ancestor of President Dwight D. Eisenhower, emigrates from the German Palatinate.

1742
Christopher Saur, a printer in Germantown, Pennsylvania, publishes the first Bible in a European language in the New World.

1759
Michael Hillegas opens the first music store in the United States, in Philadelphia.

1767
David Rittenhouse of Germantown builds the first planetarium in the Americas.

1778
Baron Friedrich Wilhelm von Steuben takes over the training of George Washington's Continental army.

1783
John Jacob Astor of Waldorf in Baden arrives in New York.

1825
Karl Beck helps to found the first gymnastics center in the United States.

1835
The first German American *Gesangverein* (choral society) is founded in Philadelphia.

1844
Prince Carl von Solms-Braunfels brings the first German settlers to Texas; the following year New Braunfels is founded.

1850
Levi Strauss, son of a Bavarian clock maker, arrives in San Francisco with bolts of heavy canvas cloth. He makes pants from the cloth, called denim, and sells them to miners.

1853
Heinrich Steinweg opens a piano factory in New York, called Steinway and Sons.

1856
Margaretha Meyer Schurz opens a kindergarten—reputed to be the first in the United States—in Watertown, Wisconsin.

1877
Carl Schurz becomes secretary of the interior.

1878
Leopold Damrosch founds the New York Symphony Society.

1883
Washington Augustus Roebling supervises completion of the Brooklyn Bridge, designed and begun by his father Johann Augustus Roebling.

1884
Ottmar Mergenthaler invents the Linotype machine for setting type, revolutionizing the printing industry.

1890
Henry Villard, born in the Palatinate, forms the General Electric Company.

1917
The United States enters World War I by declaring war on Germany.

1933
Adolf Hitler takes control of the German government, causing refugees to flee to the United States.

1935
Senator Robert F. Wagner of New York introduces the Labor Relations Act, which ensured the right of labor unions to bargain collectively with employers.

1939
Albert Einstein informs President Franklin Roosevelt that it is possible to build an atomic bomb.

1941
The United States enters World War II against Germany, Japan, and Italy.

1983
German Americans celebrate the tercentenary of German immigration to the United States.

FURTHER READING

General Accounts of German American History

Galicich, Anne. *The German Americans.* New York: Chelsea House, 1989.

Glaser, Wolfgang. *Americans and Germans: A Handy Reader and Reference Book.* Munich: Verlag Moos, 1986.

Luebke, Frederick C. *Germans in the New World.* Urbana: University of Illinois Press, 1990.

Multmann, Günter, ed. *Germans to America: 300 Years of Immigration 1683–1983.* Stuttgart: Institute for Foreign Cultural Relations, 1983.

O'Connor, Richard. *The German-Americans, An Informal History.* Boston: Little, Brown, 1968.

Rippley, La Vern J. *The German-Americans.* Lanham, Md.: University Press of America, 1984.

Trommler, Frank, and Joseph McVeigh, eds. *America and the Germans.* Philadelphia: University of Pennsylvania Press, 1985.

Wust, Klaus, and Heinz Moos. *Three Hundred Years of German Immigrants in North America, 1683–1983.* Baltimore, Md.: Heinz Moos, 1983.

Specific Aspects of German American History

Dobbert, Guido Andre. *The Disintegration of an Immigrant Community: The Cincinnati Germans, 1870–1920.* New York: Marno, 1990.

Hofmeister, Rudolf A. *The Germans of Chicago.* Champaign, Ill.: Stipes, 1976.

Kraybill, Donald B. *The Riddle of Amish Culture.* Baltimore, Md.: Johns Hopkins University Press, 1989.

Lich, Glen E. *The German Texans.* San Antonio: Institute of Texan Cultures, 1981.

Merrill-Mirsky, Carol. *Exiles in Paradise.* Los Angeles: Hollywood Bowl Museum, 1991.

Miller, Michael M. *Researching the Germans from Russia.* Fargo: North Dakota Institute for Regional Studies, 1987.

Nadel, Stanley. *Little Germany: Ethnicity, Religion, and Class in New York City, 1845–80.* Urbana: University of Illinois Press, 1990.

Redekop, Calvin Wall. *Mennonite Society.* Baltimore, Md.: Johns Hopkins University Press, 1989.

Tolzmann, Don Heinrich. *Cincinnati's German Heritage.* Bowie, Md.: Heritage Books, 1994.

First-Person Accounts of German American Life

Ameringer, Oscar. *If You Don't Weaken.* Norman: University of Oklahoma Press, 1983.

Bartholdt, Richard. *From Steerage to Congress.* Philadelphia: Dorrance, 1930.

Diede, Pauline Neher. *Homesteading on the Knife River Prairies.* Bismarck, N. Dak.: Germans from Russia Heritage Society, 1983.

Handlin, Mimi, and Layton, Marilyn Smith. *Let Me Hear Your Voice: Portraits of Aging Immigrant Jews.* Seattle: University of Washington Press, 1983.

Jordan, Gilbert J. *Yesterday in the Texas Hill Country.* College Station, Tex.: Texas A&M University Press, 1979.

Kamphoefner, Walter D., Wolfgang Helbich, and Ulrich Sommer, eds. *News from the Land of Freedom: German Immigrants Write Home.* Ithaca: Cornell University Press, 1991.

Marcus, Jacob Rader. *Memoirs of American Jews.* Philadelphia: Jewish Publication Society of America, 1955/5715.

Meyer, Ernest L. *Bucket Boy: a Milwaukee Legend.* New York: Hastings House, 1947.

Pankratz, Lydia. *Pioneer Days of the Alexanderwohl Community.* North Newton, Kans.: Mennonite Library and Archives, 1971.

Reichel, Sabine. *What Did You Do in the War, Daddy?* New York: Hill & Wang, 1989.

Roebling, Johann Augustus. *Diary of My Journey in the Year 1831.* Trenton, N.J.: Roebling Press, 1931.

Schroeder, Adolf E., and Carla Schulz-Geisberg, eds. *Hold Dear, As Always.* Columbia: University of Missouri Press, 1988.

Schurz, Carl. *The Autobiography of Carl Schurz.* New York: Scribner, 1967.

Stratton, Johanna L. *Pioneer Women: Voices from the Kansas Frontier.* New York: Simon & Schuster, 1981.

Trapp, Maria Augusta. *The Story of the Trapp Family Singers.* Philadelphia: Lippincott, 1949.

Welk, Lawrence, with Bernice McGeehan. *Wunnerful, Wunnerful.* Englewood Cliffs, N.J.: Prentice-Hall, 1971.

TEXT CREDITS

Main Text

p. 12, top: Anne Galicich, *The German Americans* (New York: Chelsea House, 1989), 20.

p. 12, middle: Norbert Krapf, *Finding the Grain* (Jasper, Ind.: Dubois County Historical Society, 1977), 22-23.

p. 12, bottom: Reprinted from *Hold Dear As Always: Jette, a German Immigrant Life in Letters*, edited by Adolf E. Schroeder and Carla Schulz-Geisberg, by permission of the University of Missouri Press and Carla Schulz-Geisberg. Copyright © 1988 by the Curators of the University of Missouri, page 34.

p. 13: Carl Schurz, *The Autobiography of Carl Schurz* (New York: Scribners, 1967), 3-5.

p. 14: From *If You Don't Weaken: The Autobiography of Oscar Ameringer*, by Oscar Ameringer. Copyright © 1983 by the University of Oklahoma Press, 36-37.

p. 15: Hamilton Holt, ed., *The Life Stories of Undistinguished Americans* (New York: Routledge, 1990), 77-79.

p. 16: Walter Hoops interview, University of Missouri—St. Louis, Archive and Manuscript Division, 1971, page 15.

p. 17, top: John Otto Reinemann, *Carried Away: Recollections and Reflections* (Philadelphia: privately published, 1976), 15-16.

p. 17, bottom: Mimi Handlin and Marilyn Smith Layton, *Let Me Hear Your Voice: Portraits of Aging Immigrant Jews* (Seattle: University of Washington Press, 1983), 60.

p. 18, top: Shirley Fischer Arends, *The Central Dakota Germans: Their History, Language and Culture* (Washington, D.C.: Georgetown University Press, 1989), 55.

p. 18, bottom: "The Life and Experiences of Sarah Harder Warkentin," *Journal of the American Historical Society of Germans from Russia* 8, no. 4 (Winter 1985): 41.

p. 19: Pauline Neher Diede, *Homesteading on the Knife River Prairies* (Bismarck, N.Dak.: Germans from Russia Heritage Society, 1983), 12.

p. 20: Sister Reinhardt Hecker interview (Bismarck, N. Dak.: Historical Data Project, 1992), 17-19.

p. 26, top: "Hooray, We're Going to America," by Ernst Bohning, edited by Jill Carter Knuth, 1063 Vernier Place, Stanford, CA 94305, page 2.

p. 26, bottom: Glen E. Lich, *The German Texans* (San Antonio: Institute of Texan Cultures, 1981), 16.

p. 27, top: Richard Bartholdt, *From Steerage to Congress* (Philadelphia: Dorrance, 1930), 13-14.

p. 27, middle: Holt, *Life Stories*, 80-81.

p. 27, bottom: George Grosz, *A Little Yes and a Big No: The Autobiography of George Grosz* (New York: Dial, 1946), 264.

p. 28, top: Theodor Engelmann, *Reminiscences*, Missouri Historical Society Archives, St. Louis.

p. 28, bottom: Reprinted from *Hold Dear As Always: Jette, a German Immigrant Life in Letters*, edited by Adolf E. Schroeder and Carla Schulz-Geisberg, by permission of the University of Missouri Press and Carla Schulz-Geisberg. Copyright © 1988 by the Curators of the University of Missouri, page 62.

p. 29, top: Stanley E. Voth, comp., *Cornelius Voth and Helena Richert Voth Family Tree* (Goessel, Kans.: Mennonite Heritage Museum, 1979), 4.

p. 29, bottom: Emil Mallinckrodt Papers, Missouri Historical Society Archives, St. Louis.

p. 30: "Hooray, We're Going to America," by Ernst Bohning, edited by Jill Carter Knuth, 1063 Vernier Place, Stanford, CA 94305, 4.

p. 31, top: Joanna L. Stratton, *Pioneer Women: Voices from the Kansas Frontier* (New York: Simon & Schuster, 1981), 15.

p. 31, bottom: Handlin and Layton, *Portraits*, 18.

p. 32, top: Henry Steele Commager and Allan Nevins, eds., *The Heritage of America* (Boston: Little, Brown, 1949), 76.

p. 32, middle: Johann Augustus Roebling, *Diary of My Journey in the Year 1831* (Trenton, N.J.: Roebling Press, 1931), 87.

p. 32, bottom: Lich, *The German Texans*, 77.

p. 33: Walter D. Kamphoefner, Wolfgang Helbich, and Ulrich Sommer, eds., *News from the Land of Freedom: German Immigrants Write Home* (Ithaca, N.Y.: Cornell University Press, 1991), 371.

p. 34: Kamphoefner et al., *News from the Land of Freedom*, 409-411.

p. 35, top: William C. Sherman et al., eds., *Plains Folk: North Dakota's Ethnic History* (Fargo: North Dakota Institute for Regional Studies at North Dakota State University, 1988), 133.

p. 35, bottom: Pauline Neher Diede, *Homesteading on the Knife River Prairies* (Bismarck, N. Dak.: Germans from Russia Heritage Society, 1983), 16.

p. 38, top: Francis Daniel Pastorius, "German Settlers in Pennsylvania" in *The Annals of America*, vol. 1 (Chicago: Encyclopedia Britannica, 1968), 311.

p. 38, bottom: Lucy Forney Bittinger, *The Germans in Colonial Times* (New York: Russell & Russell, 1901), 187-88.

p. 39, top: George Fenwick Jones, ed., *Detailed Reports on the Sulzberger Emigrants Who Settled in America* (Athens: University of Georgia Press, 1968), 187-88.

p. 39, bottom: Commager and Nevins, *Heritage*, 78-79.

p. 40: Klaus Wust and Heinz Moos, eds., *Three Hundred Years of German Immigrants in North America, 1683–1983* (Baltimore: Heinz Moos, 1983), 61.

p. 41, top: Wust and Moos, *Three Hundred Years*, 61.

p. 41, middle: Dieter Cunz, *The Maryland Germans* (Princeton, N.J.: Princeton University Press, 1948), 169.

p. 41, bottom: Galicich, *The German Americans*, 42-43.

p. 42, top: Roebling, *Diary*, 102.

p. 42, bottom: Englemann, *Reminiscences*, 41.

p. 43, top: Reprinted from *Hold Dear As Always: Jette, a German Immigrant Life in Letters*, edited by Adolf E. Schroeder and Carla Schulz-Geisberg, by permission of the University of Missouri Press and Carla Schulz-Geisberg. Copyright © 1988 by the Curators of the University of Missouri, page 68.

p. 43, middle: Jeffrey A. Hess, *Three Immigrant Stories* (Kalina, Iowa: Mennonite Historical Society, 1977), 15.

p. 43, bottom: Kamphoefner et al., *News from the Land of Freedom*, 410-11.

p. 44, top: Allan O. Kownslar, *The Texans: Their Land and History* (New York: American Heritage, 1972), 150-51.

p. 44, bottom: Reprinted from *Hold Dear As Always: Jette, a German Immigrant Life in Letters*, edited by Adolf E. Schroeder and Carla Schulz-Geisberg, by permission of the University of Missouri Press and Carla Schulz-Geisberg. Copyright © 1988 by the Curators of the University of Missouri, pages 69-70.

p. 45: "Hooray, We're Going to America," by Ernst Bohning, edited by Jill Carter Knuth, 1063 Vernier Place, Stanford, CA 94305, page 6.

p. 46: Cornelius Duerksen, *Day Book of Cornelius Duerksen* (North Newton, Kans.: Mennonite Library and Archives, undated), 7-8.

p. 47: Sophia Kallenberger Beck interview (Bismarck, N. Dak.: Historical Data Project, 1939), 3.

p. 48, top: Roebling, *Diary*, 110.

p. 48, middle: Herman Seele, *The Cypress, and Other Writings of a German Pioneer in Texas*, trans. Edward C. Breitenkamp (Austin: University of Texas Press, 1979), 47.

p. 48, bottom: Alphabetical Files (Gottlieb and Dorothea Klein), Missouri Historical Society Archives, St. Louis.

p. 49, top: Hartmut Keil and John B. Jenz, eds., *German Workers in Industrial Chicago, 1850–1910* (DeKalb: Northern Illinois University Press, 1983), 153.

p. 49, middle: Sister Reinhardt Hecker interview (Bismarck, ND: Historical Data Project, 1992), 29.

p. 49, bottom: Maria Augusta Trapp, *The Story of the Trapp Family Singers* (Philadelphia: Lippincott, 1949), 132-34.

p. 54: Kamphoefner et al., *News from the Land of Freedom*, 197.

p. 55, top: Kamphoefner et al., *News from the Land of Freedom*, 265.

p. 55, middle: Sherman et al., *Plains Folk*, 138.

p. 55, bottom: Stratton, *Pioneer Women* 104-5.

p. 56: Sophia Kallenberger Beck interview (Bismarck, N. Dak.: Historical Data Project, 1939), 3-5.

p. 57: Lydia Pankratz, *Pioneer Days of the Alexanderwohl Community* (North Newton, Kans.: Mennonite Library and Archives, 1971), 6-8.

p. 58, top: Pauline Neher Diede, *Homesteading on the Knife River Prairies* (Bismarck, N. Dak.: Germans from Russia Heritage Society, 1983), 22-23.

p. 58, bottom: Lawrence Welk with Bernice McGeehan, *Wunnerful, Wunnerful: The Autobiography of Lawrence Welk* (Englewood Cliffs, N.J.: Prentice-Hall, 1971), 4, 6-7.

p. 59, top: Kownslar, *The Texans*, 155.

p. 59, bottom: Lich, *The German Texans*, 48.

p. 61: Gilbert J. Jordan, *Yesterday in the Texas Hill Country* (College Station: Texas A&M University Press, 1979), 30-31.

p. 62, top: Doyce B. Nunis, Jr., ed., *The Golden Frontier: Recollections of Herman Francis Reinhart, 1851–1869* (Austin: University of Texas Press, 1962), 62.

p. 62, bottom: Jacob Rader Marcus, *Memoirs of American Jews*, vol. 2 (Philadelphia: Jewish Publication Society of America, 1955/5715), 42.

p. 63: Kamphoefner et al., *News from the Land of Freedom*, 417.

p. 64, top: Hartmut Keil and John B. Jenz, *German Workers in Chicago: A Documentary History of Working-Class Culture from 1850 to World War I* (Urbana: University of Illinois Press, 1988), 25.

p. 64, bottom: Kamphoefner et al., *News from the Land of Freedom*, 333.

p. 65, top: Ameringer, *If You Don't Weaken*, Copyright © 1983 by the University of Oklahoma Press, 44.

p. 65, bottom: Keil and Jenz, *German Workers in Chicago*, 59-60.

p. 66, top: Keil and Jenz, *German Workers in Chicago*, 94.

p. 66, bottom: Ernest L. Meyer, *Bucket Boy: A Milwaukee Legend* (New York: Hastings House, 1947), 1-3.

p. 67: Reprinted from *American Mosaic: The Immigrant Experience in the Words of Those Who Lived It*, by Joan Morrison and Charlotte Fox Zabusky, © 1980, 1993, by permission of the University of Pittsburgh Press, 71-72.

p. 68, top: Kamphoefner et al., *News from the Land of Freedom*, 343.

p. 68, bottom: Kamphoefner et al., *News from the Land of Freedom*, 389-90.

p. 69, top: Keil and Jenz, *German Workers in Chicago*, 78-79.

p. 69, bottom: Keil and Jenz, *German Workers in Chicago*, 82-83.

p. 70: Keil and Jenz, *German Workers in Chicago*, 89-91.

p. 72, top: Helen Campbell, *Prisoners of Poverty* (Boston: Roberts Brothers, 1890), 106-7.

p. 72, bottom: Holt, *Life Stories*, 81-82, 86.

p. 73: Giles R. Wright, *Looking Back: Eleven Life Histories* (Trenton, N.J.: New Jersey Historical Commission, 1986), 26.

PICTURE CREDITS

p. 74: Ameringer, *If You Don't Weaken,* Copyright © 1983 by the University of Oklahoma Press, 44-45.

p. 75: Keil and Jenz, *German Workers in Chicago,* 296.

p. 80, top: Schurz, *Autobiography,* 126-27.

p. 80, bottom: Meyer, *Bucket Boy,* 213, 219-23.

p. 83, top: Ameringer, *If You Don't Weaken,* Copyright © 1983 by the University of Oklahoma Press, 71.

p. 83, bottom: Bartholdt, *From Steerage to Congress,* 83.

p. 84: Richard O'Connor, *The German-Americans: An Informal History* (Boston: Little, Brown, 1968), 293.

p. 85, top: Keil and Jenz, *German Workers in Chicago,* 210.

p. 85, bottom: Excerpts from *You Must Remember This: An Oral History of Manhattan from the 1890s to World War II,* copyright © 1989 by Jeff Kisseloff, reprinted by permission of Harcourt Brace & Company, 103-5.

p. 87, top: Lich, *The German Texans,* 145-146.

p. 87, bottom: Elsie Mohr interview (Morris, Minn.: West Central Minnesota Historical Center, 1974), 2, 8, 10.

p. 88: Welk, *Wunnerful, Wunnerful,* 49.

p. 90, top: Rudolf A. Hofmeister, *The Germans of Chicago* (Champaign: Stipes, 1976), 20-21.

p. 90, bottom: Lich, *The German Texans,* 34.

p. 91: Meyer, *Bucket Boy,* 90-93.

p. 92, top: Marcus, *Memoirs,* 279.

p. 92, bottom: Sally Roesch Wagner, ed., *Daughters of Dakota,* vol. 4 (Yankton, S. Dak.: Daughters of Dakota, 1991), 6.

p. 93, top: O'Connor, *The German-Americans,* 113, 115.

p. 93, bottom: Hermann Hagedorn, *The Hyphenated Family: An American Saga* (New York: Macmillan, 1960), 81-82.

p. 94: Welk, *Wunnerful, Wunnerful,* 25-28.

p. 95: Cecilia Kost Butala, interviewed by Rose Mary Lang, Stearns County Historical Society archives, tape no. 576, 4/5/1978.

p. 96, top: Pankratz, *Pioneer Days,* 11-12.

p. 96, bottom: Hagedorn, *The Hyphenated Family,* 91.

p. 97: Marcus, *Memoirs,* 273.

p. 98, top: Kamphoefner et al., *News from the Land of Freedom,* 380-81.

p. 98, middle: Jordan, *Yesterday in the Texas Hill Country,* 83-85.

p. 98, bottom: Edna M. Boardman, "The Beautiful Stream: A Reminiscence of the Mennonite Brethren Faith" in *Heritage Review* 15, no. 2 (April 1985): 329.

p. 100, top: Henry Villard, *Memoirs,* vol. 1 (New York: Da Capo, 1969), 82-84.

p. 100, bottom: Lich, *The German Texans,* 141-42.

p. 101: Meyer, *Bucket Boy,* 227-28.

p. 102, top: Giles R. Wright with Howard L. Green and Lee R. Parks, *Schooling and Education* (Trenton: New Jersey Historical Commission, 1987), 32.

p. 102, bottom: Alfred H. Kleine-Kreuzmann, personal interview, 12/11/94.

p. 103: Mary Conrad Van Grinsven, *Little Red School House* (St. Cloud, Minn.: Stearns County Historical Society, n.d.), 1-3.

p. 104: Hagedorn, *The Hyphenated Family,* 222-23.

p. 105, top: Excerpts from *You Must Remember This: An Oral History of Manhattan from the 1890s to World War II,* copyright © 1989 by Jeff Kisseloff, reprinted by permission of Harcourt Brace & Company, 118.

p. 105, bottom: Elsie Mohr interview, 4.

p. 106, top: Ameringer, *If You Don't Weaken,* Copyright © 1983 by the University of Oklahoma Press, 329-30.

p. 106, bottom: H. Norman Schwarzkopf, *It Doesn't Take a Hero* (New York: Bantam, 1992), 8-10.

p. 107: Reprinted from *American Mosaic: The Immigrant Experience in the Words of Those Who Lived It,* by Joan Morrison and Charlotte Fox Zabusky, ©1980, 1993, by permission of the University of Pittsburgh Press, 73-74.

p. 112, top: Kurt Vonnegut, *Fates Worse Than Death* (New York: Putnam, 1991), 199-200.

p. 112, bottom: Alfred H. Kleine-Kreuzmann, personal interview, 12/11/94.

p. 113: Michael Miller, personal interview, 9/7/95.

p. 114: Hagedorn, *The Hyphenated Family,* 85-88.

p. 115: Jordan, *Yesterday in the Texas Hill Country,* 123.

p. 116: Rev. Bill Eckhardt, personal interview, 10/4/95.

Sidebars

p. 17: Eric H. Cornell, *The Lord Is My Shepherd,* unpublished manuscript, 18-19.

p. 27: Sabine Reichel, *What Did You Do in the War, Daddy?* (New York: Hill & Wang, 1989), 192-93.

p. 44: Gilbert J. Jordan, *Yesterday in the Texas Hill Country* (College Station: Texas A&M University Press, 1979), 34.

p. 46: Alberta Eiseman, *From Many Lands* (New York: Atheneum, 1974), 61.

p. 65: Stanley Nadel, *Little Germany: Ethnicity, Religion, and Class in New York City, 1845–1980* (Urbana: University of Illinois Press, 1990), 71.

p. 75: Nadel, *Little Germany,* 145.

p. 81: Nadel, *Little Germany,* 140-41.

p. 83: Nadel, *Little Germany,* 105-6.

p. 84: Nadel, *Little Germany,* 36.

p. 85: Nadel, *Little Germany,* 107.

p. 89: Nadel, *Little Germany,* 36.

p. 112: Reichel, *What Did You Do in the War, Daddy?,* 213-14.

American Historical Society of Germans from Russia: 18, 19 bottom, 20 top, 43 bottom; American Jewish Archives: 39 bottom; Arizona Historical Society: 60; Hauberg Collection, Special Collections, Augustana College Library, Rock Island, Ill.: 57 top; Beiser Family Photographs, Balch Institute for Ethnic Studies: 10; Bildarchiv Preussischer Kulturbesitz, Berlin: 14 top; Bukovina Society: 58 top; Cavalier County Historical Society, N. Dak.: 59; Chicago Historical Society: 68 top (#ICHi-04076), 75 (#ICHi-17258, A. Wittemann); Cincinnati Historical Society: 65 (#B-84-131), 72 top (#116-42D-4987), 89, 100, 101 top (#B-83-131); Circle in the Square Theatre: 7 middle; courtesy of the Colorado Historical Society: 83 right; by courtesy of the Ellis Island Immigration Museum: 49; Arthur Flegel, 19 top, 20 bottom, 27, 46, 62 top; Fort Hays Historical Society, Kans.: 54 top, 96 bottom, 111; Frankenmuth Historical Museum, Mich.: 91 top, 114, 115 top, 117; Fred Hultstrand History in Pictures Collection, NDIRS-NDSU, Fargo, N. Dak.: 54 bottom; courtesy of the George Eastman House: 69 top; German Information Service, New York: 24, 34, 38, 39 top, 40 bottom, 48 top and bottom, 103; German Society of Pennsylvania: 12 bottom, 13 top; Greater Cincinnati Convention and Visitors Bureau: 115 bottom; Marvin and Marion Hartmann: 21; courtesy of H. J. Heinz Company: 69 bottom; Idaho Historical Society: 61 top (#73-59.2); Immigrant City Archives: 29 bottom, 50, 73 top, 81, 82 top, 82 bottom, 85, 99 bottom; Indiana University/Purdue University at Indianapolis: 88 top, 106 top; Institute of Texan Cultures: 8, 25; Kansas Collection, University of Kansas Libraries: 80 bottom, 93 bottom, 96 bottom; Kansas State Historical Society: 104; Larry Keighley, Blue Bell, Pa.: 121 top; courtesy of Werner Klemperer: 6, 7 top and bottom; Victor Knell: 98 top; courtesy of the Leo Baeck Institute, New York: 11, 14 bottom; Lewis W. Hine Collection, U.S. History, Local History and Genealogy Division, the New York Public Library, Astor, Lenox and Tilden Foundations: 71 top; Library of Congress: 15 top, 15 bottom, 26 top, 28, 35, 67 bottom, 102 top, 107; Library of the Daughters of the Republic of Texas at the Alamo: cover; courtesy of Dr. Paul L. Maier: 118, 119, 120, 121 bottom; Mennonite Historical Society of Iowa: 55; Michael Miller: 112; courtesy of the Milwaukee County Historical Society: 32, 71 bottom, 72 bottom, 76, 84, 86, 90, 92 bottom, 97; Milwaukee Public Library: 53, 66, 67 top; Minnesota Historical Society: frontispiece, 45, 83 left (Ingersoll photo), 105 top; Missouri Historical Society, St. Louis: 78 (World's Fair #188), 91 bottom (Groups #341); courtesy of the Murphy Library, University of Wisconsin—La Crosse: 93 top, 105 bottom, 110, 115 bottom; Museum of the City of New York, Byron Collection: 64 top, 87 bottom; National Archives: 22, 30, 33; National Council of Jewish Women: 73 bottom; New-York Historical Society: 40 top, 44; New York Public Library Picture Collection: 16, 41, 47, 74; New York State Library: 70; Ohlinger: 31, 113; courtesy of Paterson (N.J.) Museum: 36; The Peale Museum, Baltimore City Life Museums: 29 top, 42, 92 top, 102 bottom; Antonio Perez: 108; Provincial Motherhouse of the Sisters of St. Francis, Hankinson, N. Dak.: 99 top; Putnam Museum of History and Natural Science, Davenport, Iowa: 17; Security Pacific National Bank Photograph Collection/Los Angeles Public Library: 88 bottom; courtesy of the Sophienburg Museum and Archives, New Braunfels, Tex.: 63, 64 top, 87 top, 106 bottom; State Historical Society of North Dakota: 52; State Historical Society of Wisconsin: 26 bottom (WHi x3 47140), 56 (WHi x3 44075), 58 bottom (WHi K91 410), 61 bottom (WHi A62 6784), 68 bottom (WHi IHAA 1276), 80 top (WHi x3 42083); Staten Island Historical Society: 43 bottom; from the collection of the Stearns County Historical Society, St. Cloud, Minn.: 57 bottom, 95 top and bottom; Temple Beth El, Providence, R.I.: 98 bottom; Western Michigan University Archives and Regional History Collection: 101 bottom.

INDEX

ACKNOWLEDGMENTS

We owe special thanks for the generous help we received from Michael Miller, curator of the Germans from Russia Heritage Collection; Arthur E. Flegel; Eartha Dengler and Ken Skulski of the Immigrant City Archives; Dr. Don Heinrich Tolzmann; Alfred H. Kleine-Kreuzmann; Pauline Neher Diede; Rev. Bill Eckhardt; Carla Schulz-Geisberg; and Larry Welk.

We also want to acknowledge the contributions of Thomas F. Altmann of the Milwaukee Public Library; Linda Bailey of the Cincinnati Historical Society; Judy Belan of the Augustana College Library; David Benjamin and Angela V. Currie of the State Historical Society of Wisconsin; Deborah Brown of the Missouri Historical Society; Jürgen Büssenschütt of Förderverein Deutsches Auswanderermuseum; Diane Bruce of the Institute of Texan Cultures; John E. Bye of the Dakota Institute for Regional Studies; Carolyn Cole of the Los Angeles Public Library; Ellen Crain of the Butte-Silver Bow Public Archives; Tara Deal and Nancy Toff, our hardworking and talented editors; John Decker of the Stearns County Historical Society; Carlotta de Fillo of the Staten Island Historical Society; Jeffrey S. Dosik of the National Park Service; Gail E. Farr of the Balch Institute for Ethnic Studies; Zelli Fishchetti of the Western Historical Manuscript Collection, University of Missouri at St. Louis; Professor Henry Geitz of the Max Kade Institute; Sister M. Alice Grommesh, OSF; Lois Gugel of the Mennonite Historical Society; Hans Hachmann of the Lieder-kranz Foundation; Rev. Marvin Hartmann; Nikki Heller of the National Council of Jewish Women; Barbara Hoehnle of the Amana Heritage Society; Sue Husband of Western Michigan University Archives; Jane Jackson and Iris Schumann of the Sophienburg Museum and Archives; Victor Knell; Kathy Lafferty of the Kansas Collection, University of Kansas Libraries; Rebecca Lintz of the Colorado Historical Society; Dr. Hans-Dieter Loose of Staatsarchiv Hamburg; Janice Louwagie of the Southwest Minnesota Historical Center; Mary Markey of The Peale Museum; Kristine Nelson of the West Central Minnesota Historical Center; Mary Neuchterlein of the Frankenmuth Historical Association; Dr. Margarethe Neumann of the German Society of Philadelphia; Steve Nielson and Bonnie Wilson of the Minnesota Historical Society; Wataru E. Okada of the George Eastman House; Antonio Perez; Tony Pisani of the Museum of the City of New York; Dr. Karl Heinz Pütz of the Bildarchiv Preussischer Kulturbesitz; Eunice J. Schlichting of the Putnam Museum; Phyllis Schmidt and Jerry Braun of the Fort Hays State University Center for Ethnic Studies; Kris Schmucker of the Mennonite Heritage Museum; Cathy Schultz of the American Historical Society of Germans from Russia; Judith Simonsen of the Milwaukee County Historical Society; Janice L. Smiley of the University of Missouri Press; Dr. Diane Spielmann of the Leo Baeck Institute; Christie Stanley of the Kansas State Historical Society; Todd Strand of the State Historical Society of North Dakota; Gail Stull of the H. J. Heinz Company; Fanny Valentine of the Cavalier County Historical Society; Cynthia Weisinger of the Murphy Library, University of Wisconsin at La Crosse; Oren Windholz of the Bukovina Society; Linda Ziemer of the Chicago Historical Society.

Finally, we owe a personal debt of gratitude to Dr. Paul L. Maier for opening his own German American family album to us and sharing his recollections.

ABOUT THE AUTHORS

Dorothy and Thomas Hoobler have published more than 60 books for children and young adults, including *Margaret Mead: A Life in Science; Vietnam: Why We Fought; Showa: The Age of Hirohito; Buddhism;* and *Japanese Portraits.* Their works have been honored by the Society for School Librarians International, the Library of Congress, the New York Public Library, the National Council for Social Studies, and *Best Books for Children,* among other organizations and publications. The Hooblers have also written several volumes of historical fiction for children, including *Next Stop Freedom, Frontier Diary, The Summer of Dreams,* and *Treasure in the Stream.* Dorothy Hoobler received her master's degree in American history from New York University and worked as a textbook editor before becoming a full-time free-lance editor and writer. Thomas Hoobler received his master's degree in education from Xavier University and has worked as a teacher and textbook editor.